T0296154

CAMBRIDGE COMPARATIVE PHYSIOLOGY

BIOLOGICAL CHEMISTRY AND PHYSICS OF SEA WATER

BIOLOGICAL CHEMISTRY AND PHYSICS OF SEA WATER

BY

H. W. HARVEY, M.A.

HYDROGRAPHER AT THE LABORATORY
OF THE MARINE BIOLOGICAL
ASSOCIATION, PLYMOUTH

CAMBRIDGE
AT THE UNIVERSITY PRESS
1928

CAMBRIDGE
UNIVERSITY PRESS

University Printing House, Cambridge CB2 8BS, United Kingdom

Cambridge University Press is part of the University of Cambridge.

It furthers the University's mission by disseminating knowledge in the pursuit of education, learning and research at the highest international levels of excellence.

www.cambridge.org
Information on this title: www.cambridge.org/9781107502512

First published 1928
First paperback edition 2015

A catalogue record for this publication is available from the British Library

ISBN 978-1-107-50251-2 Paperback

To

ALL THOSE COLD BLOODED ANIMALS
WHO LIVE IN THE SEA

PREFACE

During the years immediately before and since the war, much information has been collected concerning variations in the constitution and physical properties of water in the seas.

Many of these recent advances in Physical Oceanography have not been correlated nor reviewed from the point of view of the biologist. From the publication in 1912 of *The Depths of the Ocean*, a comprehensive account of Oceanography by several contributors, a gap exists until 1926. The *Journal du Conseil International pour l'Eexploration de la Mer* then commenced to publish, every three months, a bibliography of current literature, general articles, and reviews of original papers having a direct bearing upon life in the sea.

It is hoped that the information and references to original papers brought together in this volume may to some extent bridge this gap, and also be of assistance to zoologists and physiologists engaged in the study of organisms living in sea water. In addition to fishery research and the ecological study of marine life, increasing attention is being paid to the attack of various physiological problems by means of experiments with simple marine animals, rather than with animals possessing the peculiar adaptations necessary for life in fresh water or on land.

I am indebted to numerous friends and colleagues for assistance. The draft and proofs have been read in whole or in part by Dr E. J. Allen, F.R.S., Dr W. R. G. Atkins, F.R.S., Mr J. N. Carruthers, M.Sc., Lieut.-Commander J. R. Lumby, R.N., Mr F. A. Potts, M.A., Mr F. S. Russell, D.S.C., B.A., Mr J. T. Saunders, M.A., and Dr C. M. Yonge, to whose numerous suggestions the value of the book is due, their help having led to the elimination of errors and to the opinions expressed being at least reasonable.

H. W. HARVEY

PLYMOUTH
1927

ERRATUM

The title of fig. 30 should read:

"Anticyclonic current formed around a sinking centre in the southern hemisphere."

CONTENTS

Chap. I. INTRODUCTION *page* 1
Conditions affecting the growth of marine plants 5
Artificial culture of diatoms 18
Conditions affecting the growth of marine animals 19
Bibliography 32

II. CHEMISTRY OF SEA WATER 36
The saline constituents 36
Gases in solution 54
Alkalinity, carbon dioxide pressure and hydrogen ion concentration 63
Determination of hydrogen ion concentration 76
Biological effects of hydrogen ion concentration 79
Sea water as a chemical medium for plant and animal life 80
Bibliography 81

III. WATER MOVEMENTS 84
Tides 84
Tidal streams 94
Ocean currents 96
Forces derived from wind 101
Archimedean forces in the sea 110
Currents in the North Atlantic 114
Submarine waves and Seiches 120
Bjerknes' circulation theory and its application 122
Bibliography 130

IV. TEMPERATURE OF THE SEA 133
Gain and loss of heat 133
Distribution of temperature with depth 139
Surface temperature 144
Influence of currents 145
Temperature of adjoining or partially enclosed seas 149
Fluctuations of temperature and currents 152
Bibliography 154

Chap. V. COLOUR AND ILLUMINATION OF SEA WATER *page* 155
Bibliography 163

VI. CHEMICAL AND PHYSICAL FACTORS CONTROLLING THE DENSITY OF POPULATION 164
Growth and consumption of algae 164
Cycle of life in the sea 167
Upwelling and vertical mixing 168
Fertility of coastal areas 173
Fertility and latitude 176
Regeneration of phosphates and nitrates 179
Fluctuations in Fertility 181
Bibliography 188

INDEX 189

Chapter I

INTRODUCTION

The changes in appearance of the sea on passing from one region to another, and on approaching land, have been of interest and consequence since the earliest ages. It was largely a study of such phenomena which helped the Norwegians to make fishing voyages as far as Iceland and Greenland and to touch the American continent in the tenth century, without the assistance of the compass to give them direction when the stars and sun were obscured by clouds or fog. In the fifteenth century the Atlantic was again crossed by Columbus, seaman and scholar, who made full use of mathematical and astronomical knowledge obtained at the University of Pavia and the School of Cartography at Lisbon; his letters and accounts show that he was not only an astronomer but a keen observer of all natural phenomena in any way connected with the sea and placed considerable value upon such knowledge.

Up to the middle of the eighteenth century the coasts and seas were represented by rough charts evolved from imperfect reckonings made during the course of ordinary voyages. Captain Cook was among the first to be employed in making accurate charts of the coastline, depths and currents. He was engaged in surveying on the Labrador Coast previous to his appointment to the famous exploring expedition to the South Seas in 1768, and it speaks well for the accuracy and the practical value of the work that his charts of this region had a regular sale as recently as 50 years ago. The survey of the coastlines, the depths and the surface currents of the oceans and other conditions affecting shipping is now carried on by the navies of the principal governments. This constitutes the science of Hydrography as ordinarily understood. It was not until about the middle of last century that any systematic attempts were made to explore the physical and biological conditions of the oceans as a whole. Since the scientific expedition of H.M.S. *Challenger* from 1872 to 1876, under the direction of Wyville Thomson, at that time Professor of Natural History at Edinburgh, many investigations have been carried out concerning the physical and chemical conditions of the sea which affect, and even control, the plant and animal life within it. It is with these particular physical and chemical conditions and their change with place and time that this volume deals, more particularly with research subsequent to the publication of Krummel's *Handbuch der Oceanographie* in 1911. It is the branch of Hydrography which concerns the living organisms within the sea rather than the ships upon it.

At first the object of the investigations on the physical conditions of the ocean waters, mostly by the Scandinavians, seems to have been to explain and forecast variability in the general weather conditions of some parts of North-west Europe. For instance, Baltic seaports do not freeze up at exactly the same time each year, and it would be a matter of practical value if the times of freezing and opening could be forecasted. The observations of temperature and salinity of the water led to the inference that relatively warm Atlantic water slowly drifted round the north of Scotland towards the Norwegian coast and that this slow drift varied from year to year, being stronger in some years and weaker in others. A further interest was taken in these investigations since they have an obvious bearing upon the fisheries, most of which are seasonal, but do not resume at exactly the same time nor yield the same quantity of fish each year, sometimes failing altogether. It thus became apparent that knowledge such as these investigations were gradually yielding was of value to the nations on the Atlantic seaboard of Europe. An International Council was founded in 1898 to assist co-operation between investigators in the various countries and to publish observations, collected by the participating nations, and results having more than local interest. In this way a considerable amount of data, particularly relating to the surface water collected from light vessels and on steamship routes, has been made available. Ocean cruises have been made by ships solely for the purpose of investigating the biological and physical conditions of the sea, but, owing to the considerable running expenses of a vessel of suitable size to carry out the work in a satisfactory manner, the number and extent of such expeditions has been limited. It has only been a part of their programme to investigate physical and chemical conditions. Amongst these expeditions may be mentioned the following:

Voringen, in the North Atlantic, 1876–1878 (Norwegian).

Valdivia, in the Atlantic in 1878–1879, under the direction of Chun (German).

Blake and *Albatross*, in the Atlantic and Pacific in 1887–1900, under the direction of Alexander Agassiz (U.S.A.).

Yachts of the late Prince of Monaco in the Mediterranean and Atlantic.

National, plankton expedition in the North Atlantic, under the direction of Victor Hensen in 1889 (German).

Pola, in Mediterranean and Red Sea, where Natterer made numerous chemical observations in 1890–1898 (Austrian).

Ingolf, in North Atlantic and Mediterranean in 1895–1896 (Danish).

Belgica, in 1897–1898 (Belgian).

Yachts of Mr R. N. Wolfenden in the Atlantic in 1903–1907 (British).

Thor, in North-west Atlantic in 1905–1906, in Mediterranean, 1908–1910 (Danish).

Planet, in 1906 (German).

Fridjhof, in the North-west Atlantic in 1910, under the direction of Nansen (Norwegian).

Deutschland, South Polar Expedition and in the Atlantic, where Lohmann made very thorough observations of the density of minute plant life suspended in the water. 1911 (German).

Michael Sars, under the direction of Dr J. Hjort and Sir John Murray who financed the expedition, in the Atlantic. A very readable general account of physical and biological oceanography, based largely on the results of this expedition, was published by Macmillan and Co. in 1912 entitled *Depths of the Ocean* by Murray and other contributors, and forms the leading text-book of oceanographical science at the present time. 1910 (Norwegian).

Fram, under the direction of Nansen in North Polar Seas, 1893–1896 and of Amundsen, west of the British Isles in 1910 (Norwegian).

Armauer Hansen, under the leadership of B. Helland-Hansen, in the North-east Atlantic in 1913, 1914 and 1922 (Norwegian).

Veslemoy, under the direction of Nansen, in the northern waters of the North Atlantic in 1912 (Norwegian).

Scotia, in North-west Atlantic, 1913 (British).

Dana, under the direction of J. Schmidt, in the Atlantic, when definite evidence was obtained that both American and European fresh-water eels breed in the Sargasso. 1920 (Danish).

Meteor, now following a programme for a very complete physical survey of the South Atlantic drawn up by the late Dr Merz (German).

Discovery, now engaged in investigating the conditions affecting the life of whales, under the direction of Dr S. Kemp (Falkland Islands Government).

Numerous investigations of conditions near the coasts, which have thrown light upon conditions in the oceans, have been carried out at Marine Biological Stations, such as at Naples, Plymouth, Monaco and Kiel, and by vessels employed in fishery research by the European countries. Our present knowledge of Physical Oceanography is largely based on these investigations, which form the subject matter of this volume for the most part.

During recent years observations have been made on the east coast of Canada for the Canadian Government, and on the east coast of the United States the work commenced by Agassiz has been continued in the Gulf of Maine.

In addition observations are made yearly by the Atlantic Ice Patrol, a service inaugurated after the loss of the *Titanic* to warn vessels passing south of Newfoundland of the position of the larger icebergs, the drift of which the patrol follows. In this case the observations are of practical use as well as of scientific interest, for the currents, which bear the icebergs with them, may be calculated from the observed physical data.

Since it is desired to discuss the physical and chemical conditions governing life in the water rather than the conditions of particular regions, the area considered is limited to the North Atlantic and adjoining seas. Although this does not comprise more than about a tenth of the entire oceans, its features and phenomena are better known than any other from having been the great highway for several centuries and from its fisheries, the products of which have a value of about £70,000,000 annually. It owes its maritime superiority to the great proportionate length of its varied coastline and to the many rivers giving ready access and intercommunication to seats of dense and inland population. Even in this, the best known ocean, our knowledge rests upon a relatively limited number of observations scattered over a vast expanse which attains a depth of 3–4 miles over much of its area. A great deal of knowledge of the *general* conditions has been accumulated, less concerning the annual *variation* with the seasons, and relatively little concerning the *fluctuations* from year to year. It is these latter which are of the greatest interest from the point of view of the fisheries, which themselves fluctuate from year to year. One of the main objects of fishery research is to find the cause of these often considerable natural fluctuations in the numbers and size of fish inhabiting particular areas, in the hope of being able to predict them.

As an introduction to any consideration of the factors which affect living organisms, it seems desirable at the outset to describe very shortly the main types which populate the ocean waters and their more obvious requirements for a vigorous life.

The multiplicity of species of marine plants and animals precludes more than mention of the main types, sufficient however to portray the cycle of events which occur in the open sea.

MARINE PLANTS

The larger algae or seaweeds are sessile in a comparatively narrow zone around the coasts. The maximum depth at which they are found varies with the depth to which light penetrates in appreciable quantity. In addition to the fixed algae the upper layers of the sea are inhabited by a flora of very small plants, mostly unicellular, which are suspended in the water and borne along with the currents. This flora is known as the vegetable- or phyto-plankton. It is in the upper illuminated layers of the sea, extending to a depth of 20 to 150 metres, according to the clearness of the water and the amount of sunshine, that these drifting algae grow and multiply. On death or during old age they slowly sink, often with the formation of resting spores which may be carried many thousands of miles suspended in the water, which is in a slow but constant state of circulation.

The fixed algae or sea weeds are distributed along the coast in zones. Green algae occur on the upper part of the strip between high and low water mark in brackish water; on the open sea coast, if present at all, it is usually where a trickle of water flows out from the land, for, unlike the brown and red algae, some of the green algae can endure large and rapid changes in salinity. Brown algae constitute the main mass of the inter-tidal vegetation; they merge with and are gradually replaced by the red algae at greater depths. The distance to which the various classes of algae extend is dependent very markedly upon the nature of the bottom and upon the latitude. The flora of the inter-tidal zone in the tropics is limited to types which can withstand heat; on the Egyptian coast the narrow zone is nearly devoid of algal life, owing to the scorching heat, whereas around the British Isles the zone is one of luxuriant vegetation. This has been shown to continue photosynthesis very actively during the period it is left dry by the tide.

The depths to which the fixed algae penetrate varies with latitude. Owing to the lower altitude of the sun and probably also to the greater amount of sediment retained in suspension by the water of the rougher northern seas, only a sparse vegetation is found below about 35 metres. On the other hand, a fairly abundant flora of red algae has been found in the clear waters of the Mediterranean to a depth of 130 metres. In Puget Sound, Gail found the lower limit at which photosynthesis takes place by red or brown algae to be about 35 metres.

These algae afford harbourage for a population of small animals; although they are not actually eaten by fishes or invertebrates in general, the slime on their surface and the dead portions must nourish numerous bacteria and protozoa, themselves the food of rather higher animals. It has been suggested that their spores, full of nutriment, are of a size suitable as food for numerous small animals, and are given off in the neighbourhood of a luxurious vegetation in sufficient quantity to be a considerable addition to their food.

The area of the oceans in which seaweeds occur is very small indeed compared with its whole extent, since they rarely extend in quantity for a distance of more than a few hundred yards from the shore and their abundance rapidly diminishes at depths greater than a few yards below low water mark.

The small plants suspended in the water and constituting the phytoplankton are of numerous kinds, many of them so minute that they pass through a fine silk net, usually made of bolting silk, as used for sieving flour, the finest having 180 meshes to the linear inch. These minute forms, frequently termed microplankton, are obtained by centrifuging or from the digestive tracts of certain small marine animals which sieve them out of the water.

Around our coasts the phytoplankton, borne passively with the currents, is made up for the most part of diatoms—consisting of a thin film of protoplasm lining the inner wall of external silica skeletons—together with many species of a widespread group of unicellular organisms, the dinoflagellates or peridinians. Most of these have a complete external skeleton of hard substance which is practically full of protoplasm containing chromatophores. The complete skeleton precludes their preying upon animals, and they convert carbon dioxide dissolved in the sea into starch by photosynthesis through the agency of their chromatophores. Since, unlike diatoms, they are nearly full of protoplasm and also occur in considerable quantities they probably constitute a material part of the food upon which the animal community lives. Other species of the same group are not completely covered with an external skeleton and contain no chromatophores; these species live by preying on other organisms and are consequently animals and not phytoplankton.

On passing from the coastal and comparatively shallow areas over plateaux or submarine ridges into very deep water, the coastal or neritic species become much less abundant and their place is taken by oceanic species which include many more minute brown flagellate algae, protected by remarkable shields of calcium carbonate which unite into a defensive covering, than are found in coastal areas.

In the deep oceans, well away from land or shallow plateaux, the population is sparse both on the bottom and throughout the water. Hensen's Atlantic Expedition in 1889 and the more recent expeditions showed that the total quantity of both phyto- and animal-plankton in tropical and subtropical regions was considerably less than in higher latitudes. The maximum density of population occurs nearer land and over banks, such as the Grand Banks south of Newfoundland and the Faeroe Bank.

There is also a change in the plankton flora on passing from the arctic regions into temperate and tropic seas, where diatoms give way largely to blue-green algae.

Table I. Population density of microplankton organisms per litre. Lohmann (1).

Depth (metres)	Latitude				
	50°–40° N.	40°–30° N.	30°–20° N.	20°–10° N.	10° N.–equator
0	20,000	7,000	1,800	2,000	3,000
50	20,000	5,000	1,600	1,500	2,000
100	3,000	2,000	1,000	700	400
200	3,000	200	200	80	100
400	2,500	100	—	—	—
0–400	6,000 May	2,000 June	600 June	500	600

In higher latitudes quite definite periods of maximum density of particular forms of the population frequently occur; in the English Channel and Irish Sea, for instance, a burst of diatom growth takes place in the early spring and again in the autumn; in the Gulf of Maine it takes place in March, and again, but to a much less extent, in August and September. During the winter months the phytoplankton is very sparse. Plant growth is concentrated during a limited portion of the year, when it is enormously rapid compared with the continuous growth in the 'seasonless' tropics.

In the open oceans well away from land the researches of Gran, Hensen and Lohmann have all gone to show that the distribution of the oceanic phytoplankton organisms is regular. Dealing with the regularity of distribution Gran ((2) 1912) writes: "The samples from adjacent localities with similar life conditions have yielded very concordant results. I do not consider it any exception to this statement that in tropical waters dense masses of the blue-green alga, *Trichodesmium*, sometimes collect as water bloom in certain areas and not in others, or that diatoms near the edge of the polar ice occur in more or less local swarms, for I consider it more than probable that these irregularities arise because the conditions of existence vary in closely adjoining areas."

In the neighbourhood of land in our latitudes during the spring and summer months, the diatoms are not found to be evenly distributed either vertically or horizontally. A wave of production of a particular species occurs in the spring, and after reaching a maximum the diatoms will sink passively into the deeper water, to be succeeded by diminishing waves of production in the upper layers, frequently of other species. In the late summer or autumn, production increases again until limited by lack of sunlight; thus a waxing and waning shower of diatoms sinks to the bottom, many of them carrying resting spores. The following table shows the

Table II. Distribution of diatoms with depth[1].

In lat. 61° 35′ N., long. 0° 47′ E. May 25, 1912		
Depth (metres)	Temp. °C.	Diatoms per litre
0	9·0	364,520
10	—	362,680
20	—	108,440
30	—	74,800
50	—	7,920
70	—	2,480
100	—	500
200	8·17	220
In lat. 61° 38′ N., long. 0° 41′ W. May 25, 1912		
Depth (metres)	Temp. °C.	Diatoms per litre
0	9·2	87,680
10	—	74,100
20	—	14,860
30	—	14,380
50	9·03	750,200
70	—	248,160
100	—	1,460
210	8·15	2,460

[1] Gran, *Bull. Planktonique*, 1912.

distribution of diatoms with depth on the same day at two positions four miles apart.

At the first position a wave of production is taking place near the surface, at the second position production at the surface is taking place and diatoms resulting from a previous wave of production are in the layer between 50 and 70 metres, probably sinking passively.

These minute ubiquitous plants are often present in considerable quantity.

From a series of observations on the seasonal changes of alkalinity, which are due to carbon dioxide utilised during photosynthesis, Atkins (3) has made a rough estimate of the minimum annual production of phytoplankton in the English Channel as about *1400 metric tons wet weight of phytoplankton per square kilometre*. A close agreement was found between this and a subsequent estimation based on the annual change in phosphate content. Moore arrived at a slightly higher figure for the annual production in the Irish Sea from alkalinity determinations. When the greater mean depth of the Irish Sea than of the English Channel (100 metres as against 70–80 metres) is taken into consideration the agreement between the values found is extraordinarily close.

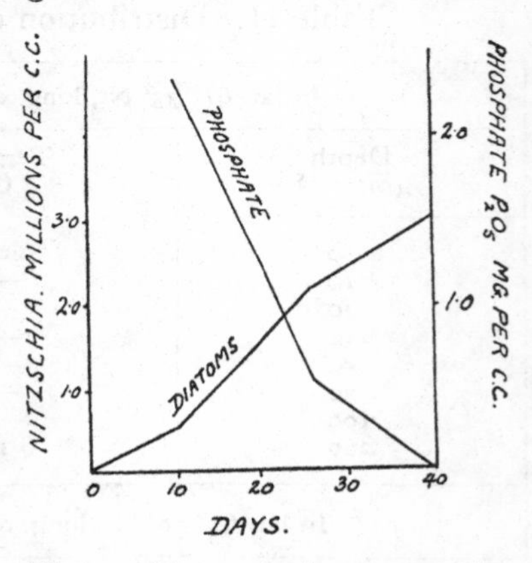

Fig. 1. Observed changes in the number of diatoms (*Nitzschia* sp.) and dissolved phosphate in a culture (Atkins).

Based on the experimental observation that when the diatom *Nitzschia closterium* multiplies in sea water enriched with nitrate and phosphate, roughly 1 gram of P_2O_5 will suffice for the production of 9×10^{11} such diatoms, a further calculation indicates that in the water of the English Channel, where at least 30 mg. P_2O_5 per cubic metre are annually used up by the phytoplankton, the minimum annual production is equivalent to 27 diatoms per cubic millimetre during the course of the year. Nothing approaching such a number of plant organisms is present in the sea simultaneously; the value relates to the production throughout the year; many of the organisms will die or be eaten after a short life and many will be much larger and equivalent to more than one *Nitzschia* diatom upon which the value is calculated. This estimate of the minimum crop in the English and Irish channels corresponds with the annual production, by photosynthesis from carbon dioxide, of 250 metric tons of glucose per square kilometre or of over 2½ lb. of vegetables below each square yard. Probably the crop is considerably greater.

Gran observed the rate at which oxygen increased in a flask of sea water, suspended in Oslo Fiord in March, due to photosynthesis by the included phytoplankton. If this rate is taken as a measure of the photosynthesis occurring in the upper ten metres of the sea, it is comparable with the formation daily of 2·4 metric tons of glucose sugar per square kilometre during the spring and autumn periods of rapid growth.

The plants in general have definite known requirements for their growth and multiplication; in addition to carbon dioxide, which occurs in plenty in sea water and is utilised in photosynthesis by the plants, they require nutrient salts, of which the phosphates and nitrates are built up into proteins. As far back as 1899 Brandt recognised the importance of these salts—never present in more than minute traces—for the growth of plant life in the sea. This surmise was borne out by the results of numerous analyses made by Raben which showed that the deep water of the open ocean contained more nitrate and phosphate than the upper layers in which phytoplankton organisms were active, and that North Sea waters contained more nitrate and phosphate in the winter than in the summer months. Brandt concluded that, with the concentration of these nutritive salts below a certain minimum value, plant life could not proceed.

Brandt also pointed out that the nitrate in the sea was continually being renewed by rain and river water, and he put forward the hypothesis that denitrifying bacteria, such as had been found in the Baltic by Bauer and in the North Sea by Feitel ((4) 1901 and 1903), reduced the concentration of nitrate, particularly in the warm surface water of the tropics, thus accounting for the smaller production of phytoplankton in these regions.

This hypothesis appears untenable since denitrifying bacteria do not break down nitrates when they have a supply of oxygen in solution. As pointed out by Gran, when grown under anaerobic conditions in a solution containing nitrate, they break it down, utilising the oxygen and setting free nitrogen, but when grown in solutions containing dissolved oxygen they do not attack the nitrates, or, at least, not to any appreciable extent ((2) pp. 369–70, 1912).

Brandt's reply to this criticism is given in "Über den Nitratgehalt des Ozeanwassers und seine biologische Bedeutung," *Abh. der Kais. Leop.-Carol. Deutschen Akad. d. Naturforscher*, **100**, No. 4, Halle, 1915.

Nathansohn recognised that ascending currents may produce a rich phytoplankton, by carrying nutritive substances up from deep water in which Raben had found them more abundant. "Where a deep-going current is pressed against the coast banks, as off the west coast of Scotland or at the Faero Bank, eddying movements, not only horizontal but also vertical will necessarily arise, and the water masses from the depths will be carried up to the surface."

Methods of analysis have recently been devised which are not subject to the errors of those used by Raben in cases where the phosphate and nitrate are at their greatest dilution (1922–1925, see pp. 40–48). These have shown that the growth of phytoplankton continues during the summer in the English Channel until the phosphates and nitrates are almost completely used up—only a few milligrams of P_2O_5 or nitrate-nitrogen remaining per cubic metre of water. A similar condition was found in the upper layers of the open ocean during summer, and throughout the year in the tropics.

The assumption (2) that the sea in the neighbourhood of land owed its greater fertility largely to nutrient salts carried out from the coast was not substantiated, since the nutrient salts were found to be very largely utilised by the plant life in these rivers during the sunny months before the water entered the sea. Their addition to the sea in temperate coastal areas by this means is altogether overshadowed by the nutrient salts re-formed yearly from dead organisms, which in these areas are restored to the upper layers by vertical circulation.

Typical results obtained and a description of the methods are given on pp. 40–48 and 168–188. They confirm the conclusions drawn by Nathansohn that ascending currents, where a deep current impinges on a submarine ridge, as between the Shetlands and Iceland, carry water rich in nutrient salts to the surface, and further show the very important part played by vertical mixing of the bottom water with the upper layers when these cool in winter, particularly in areas where the depth is moderate and cooling of the surface is considerable during the winter months.

After death the planktonic algae sink to the bottom, where by autolysis and bacterial action soluble phosphates and ammonium compounds are produced from their corpses. Ammonium ions are found in sea water and are an intermediate stage in the regeneration of nitrates.

In the course of a very thorough investigation of fresh water lakes carried out by Juday, Domogalla ((5) 1925) found that the ammonium salts, amino acids and proteins in solution in both the surface and bottom water of Lake Mendota reached a maximum in the winter, falling to a minimum in the summer, when the amount of plankton, both algal and animal, was at a maximum. In the warmer weather these compounds would be more rapidly changed by bacteria into nitrates which would in turn be more rapidly used up by plant growth, so this evidence alone does not indicate that plants and animals absorb the amino acids, etc., directly. Schreiner and Skinner ((6) 1912) found that plants grown in a nutrient solution would thrive more readily if the nutrient solution contained amino acids as well as inorganic nitrogen compounds.

Moore's experiments with fresh-water algae and with the marine alga *Enteromorpha compressus* were considered to indicate that atmospheric nitrogen was utilised during their growth, because when kept in a covered jar in sunlight their increase in combined nitrogen was found greater than the combined nitrogen initially present in the water ((7) 1921). Some subsequent experiments, notably those of Wann, confirmed Moore's opinion that algae utilised atmospheric nitrogen. These experiments were repeated and critically examined by Bristol and Page ((8) 1923) who found, without doubt, that Wann's results were based on faulty chemical procedure. In their own experiments with algae freed from bacteria no atmospheric nitrogen was utilised. They further point out that this is also the case in the more carefully controlled experiments of other workers.

Although there is no evidence that algae actually fix nitrogen, the possibility of their being able to utilise nitrogen through the agency of nitrogen-fixing bacteria must be mentioned because Reinke ((9) 1904) found nitrogen-fixing bacteria (azotobacter) on a number of algae from Heligoland, which multiplied and fixed atmospheric nitrogen when kept in a suitable culture medium. Azotobacter occurring in the slime of algae were also identified by Keutner ((10) 1905) and by Keding ((11) 1906). There is, however, no reason to suppose that azotobacter plays more than a very inconsiderable part in the economy of the open ocean—it has never been found in open sea water or in plankton.

Whether marine algae can obtain combined nitrogen from other mineral salts than nitrates is the next question which arises. Bristol and Page have shown that certain species of algae occurring in the soil ashore flourish in a medium containing no nitrate, but ammonium salts instead, under conditions which allow of no suspicion that the ammonium salts were first converted into nitrates by the agency of nitrate-forming bacteria ((8) 1923). Ammonium salts do not appear to take the place of nitrates in the sea, since they do not suffer any considerable reduction in spring and early summer when the proliferation of the phytoplankton is at a maximum and when the nitrate content is being rapidly reduced ((12) 1916–1920). If ammonium salts were utilised directly it seems almost certain that a well-marked seasonal variation would occur, just as happens with nitrates and phosphates.

The balance of evidence suggests that almost all the nitrogen built up by plants into proteins is absorbed in the form of nitrates, the amount derived from other sources being small in comparison.

In addition to limiting the total growth of plant life in any area, a lack of nutrient salts probably affects the rate of photosynthesis of those plants which are present. During an hour's sunlight an equal number of diatoms would form more carbohydrate in a sea rich in phosphate and nitrate than in a sea where these nutrient salts were deficient; this is a surmise based upon the known behaviour of higher plants ((13) 1922) but it is borne out by the observation that tropical phytoplankton organisms have fewer and paler chromatophores than nearly related northern types, giving them a distinctive appearance.

Thus in a sea with ample sunshine for phytoplankton growth, as the nutrient salts are used up, the formation of starch falls off in amount, multiplication is enormously reduced, until finally the existing plants come nearer to the stage when the rate of starch formation is exceeded by the consumption of organic matter in their own respiratory processes.

The growth of phytoplankton in temperate and higher latitudes is limited by lack of *light* during the short winter days with little sunshine. Off our coasts the water contains very little phytoplankton during the winter months. Although *continued* existence only becomes possible for plants when the light is of such an intensity and daily duration that it enables photosynthesis to be carried out at a rate at least equal to that at which the carbon compounds of the plants are dissipated through respiration, existence appears to be possible for quite a long period in the dark during which many phytoplankton algae can 'live on their own fat.' The fact that diatoms and other phytoplankton are met with alive at considerable depths is no proof that they are within the necessary limit of illumination, since they may be slowly sinking or owe their position to vertical mixing of the water.

The intensity of light at the surface together with the transparency of the water controls the thickness of the water layer in which photosynthesis takes place; it is the thickness of this layer which permits the ocean to support such an abundant flora. Phytoplankton is active throughout a considerably deeper layer in lower latitudes, and here a thicker layer is depleted of nutrient salts owing to greater intensity of illumination. The variations in transparency and the penetration of light are discussed on pages 155–163. The curves on page 160 show how rapidly light is absorbed in its passage through sea water and how the rate of absorption varies considerably in different places.

An experiment in Oslo Fiord shows that the intensity of light controls the production of plankton; surface water containing mixed phyto- and animal plankton organisms was filled into litre flasks which were hung in the Fiord at various depths. One flask protected from light by a black cloth acted as a control; in this no photosynthesis took place, and the flask lost 0·07 c.c. of oxygen in 24 hours owing to the respiration of the contained organisms. In the other flasks this loss was taking place, but at the same time oxygen was being produced by photosynthesis; in those immersed at depths of less than 10 metres the gain exceeded the loss. Organic matter was built up in the form of new phytoplankton organisms more rapidly than it was being broken down in the process of respiration by the whole community ((39) 1927).

Table III.

Depth of flasks below surface (metres)	Oxygen c.c. per litre	Oxygen produced by photosynthesis in 24 hours c.c. per litre	% increase of four species of phytoplankton organisms in the water
0	Gain 0·20	0·27	39
2	,, 0·19	0·26	55
5	,, 0·13	0·20	48
10	No change	0·07	21
20	Loss 0·03	0·04	0
30	,, 0·05	0·02	0

The experimental evidence obtained by several investigators has shown that the rate of photosynthesis is almost proportional to the light intensity, when the intensity is low.

Table IV. Influence of light intensity on the rate of photosynthesis of *Fontinalis* (Harder, *Jahrb. f. wiss. Bot.* **60**, 531–71, 1921).

Light intensity in metre candles (*a*)	Real assimilation in arbitrary units (*b*)
677	1·156
2000	3·30
6000	7·00

A considerable amount of work has been done on the quality of light which is most effective for photosynthesis. Kneip and Minder ((14) 1909) found that *Elodea* would proceed with photosynthesis actively in red and blue light, but showed little activity in green light of equal intensity, and similar results have been obtained by Wurmser on *Ulva* ((14) 1920).

Table V. Absorption of light of different wave lengths and photosynthesis by *Ulva lactuca*.

Wave length $\mu\mu$	Assimilation	Energy absorbed
750–560 (red)	100	100
560–460 (green)	24	6
460 & shorter (violet)	80	34

Klugh ((15) 1925), also working with lights of equal intensity, found that the green alga, *Volvox*, would multiply rapidly in red light, less rapidly in blue and not in green light, thus confirming previous results.

A desmid, *Closterium*, on the other hand, multiplied rapidly in red, but not in either blue or green. It is practically certain that the red algae can and do utilise light at the blue end of the spectrum, since they live at depths to which little but blue light penetrates. The capacity of the phytoplankton to utilise short wave length light is unknown.

In addition to the definite requirements for plant life already considered the *temperature* of the sea plays a part both in the distribution of the various species of phytoplankton, and in the rate at which the vital processes of all proceed. Various investigators have shown that the rate of respiration of plants increases by approximately 10 % for a rise of 1° C. (16) within their ordinary range of temperature just as it does for cold-blooded animals. The researches of Blackman and Miss Matthaei ((17) 1904) have further shown that the rate of photosynthesis, given a sufficiency of nutrient salts and light, increases at approximately the same rate with rise of temperature.

Hence in the colder seas of higher latitudes all the vital processes proceed at a slower rate and the loss of combined carbon by the living plant in the course of its respiration is less.

It is not suggested that the life processes of algae inhabiting tropical seas with a surface temperature of 20° C. actually proceed twice as fast as of those inhabiting waters of higher latitudes having a summer temperature of 10° C., since the generality of algae which populate the tropics are of different species to those in higher latitudes. However it does appear that there will be a very considerable difference in the velocity of the vital processes of the two populations.

The rate of photosynthesis and of respiration of the common brown alga *Fucus serratus* has been investigated by Kneip ((18) 1914), who found that a decrease in temperature lowers its rate of respiration to a greater extent than it lowers its rate of photosynthesis, and, further, when this alga is kept in the dark for a considerable time its rate of respiration slowly decreases without the alga perishing. If marine algae in general behave in this manner, they are particularly well adapted to life at arctic sea temperatures, 4° to − 1° C., and to survival for a considerable period in complete or nearly complete darkness.

ARTIFICIAL CULTURE OF DIATOMS

A number of diatoms and minute algae such as constitute the phytoplankton have been very successfully grown *in vitro* by Allen and Nelson ((19) 1910). The maximum healthy growth of plankton diatoms took place when the diatoms were obtained as free as possible from all other organisms, if not in a pure culture at least in a 'persistent' culture—that is, a culture in which the same diatom flourishes for a considerable time and is not crowded out by the growth of another species. Where a sample of a mixed population is cultured, such as may be obtained by towing a fine silk net in the sea (180 meshes per linear inch), the usual planktonic forms first develop in considerable numbers. Subsequently bottom diatoms and algae of various kinds become abundant, and the true plankton forms die out. It is difficult to avoid the impression that the organisms which finally take possession of the cultures are in some way directly inimical to those which they supersede, not merely by robbing them of their food supply but perhaps by the production of toxic substances. In nature, in the sea, the true planktonic forms always outnumber the other organisms which in the thick laboratory cultures invariably succeed in gaining the upper hand.

In order to obtain a maximum vigorous growth of a diatom separated from other organisms in filtered and sterilised sea water collected from the open sea, it was not sufficient merely to enrich the water with nitrate and phosphate. The further addition of a small amount of ferric chloride and calcium chloride, of which the ferric iron was completely precipitated as basic phosphate and the calcium chloride partly so, greatly enhanced the efficiency of the enriched sea water as a culture medium.

Nor did filtered and sterilised aquarium tank water, very rich in nitrates and phosphates, give a maximum vigorous growth of diatoms. Its efficiency was greatly improved by the addition of phosphate, together with ferric and calcium chloride, from which the precipitate of basic iron and calcium phosphate is produced.

This points to the tentative conclusion that sea water contains a substance or substances inhibitory to the growth of diatoms which is adsorbed upon and carried down with the precipitate. Further, the addition of hydrogen peroxide to either sterilised open sea water or sterilised aquarium tank water improved its efficiency, as did treatment with animal charcoal. It may be that these treatments remove the inhibitory substances by oxidation and adsorption respectively.

Wide ranges in total salinity of the water had no appreciable effect on the growth of the diatoms[1], which was, however, greater in strong rather than in weak diffuse light, and more satisfactory in diffuse than in direct sunlight.

[1] Legendre found that the rate at which carbon dioxide was assimilated by the estuarine seaweed, *Ulva lactuca*, was greatest in much diluted sea water, decreasing as the density of the sea water was either increased above or decreased below *ca.* 1·01. *C.R. de Soc. Biol.* **85**, 222, 1921.

An attempt to grow a typical plankton diatom *Thalassiosira gravida* in an artificial sea water prepared by dissolving the purest obtainable salts in glass distilled water led to a very surprising result ((20) 1914). No satisfactory growth could be obtained unless about 1 % of sea water was added. The result appears to be due to some specific substance present in minute quantity in the natural sea water which is essential to the vigorous growth of this diatom. Since all detectable iron is precipitated in such an artificial sea water, and iron in soluble organic combination almost certainly occurs in natural sea water, the result may possibly be due to lack of this constituent. On the other hand Peach and Drummond obtained a good growth of *Nitzschia closterium* in artificial sea water without any such addition. (*Biochem. Journ.* **18**, 464–468, 1924.)

From the foregoing summary of the conditions necessary for vigorous growth, it is apparent that plant life in the sea is only limited by the supply of phosphate and nitrate in the zone to which light of suitable wave length penetrates in sufficient quantity for the production of carbohydrate by photosynthesis in excess of the plants' immediate requirements for respiration, and by the depth of this zone.

MARINE ANIMALS

The rich vegetative growth in the upper layers supports a considerable and immensely varied animal population, *which is entirely dependent either directly or indirectly upon the algae for carbohydrates and proteins*. Some only of these animals feed on algae during the whole of their lives, others during part of their lifetime, while others are carnivorous, catching and preying upon their neighbours, or ingesting and feeding upon particles of dead organisms and bacteria.

Of these animals many live suspended in the water, having swimming power insufficient to stem the currents by which they are passively transported; they vary in size from the smallest unicellular organisms to large jelly fish, and form the animal- or zoo-plankton. Others live on the bottom or in the mud or sand, moving only short distances; in their young stages they are mainly planktonic. Yet others possess considerable swimming powers, even performing regular migrations; of these the great majority are also planktonic for a period while in the egg and larval stages.

It follows that currents exercise a powerful influence upon the fauna, particularly horizontal currents, whereas it is vertical currents which exercise the most potent effect upon the flora.

Where a current sets across an area rich in animal life, it may carry many young of fish and invertebrates to new and less suitable habitats where their chance of survival is less and their food is scarce.

An instance of considerable economic importance occurs in the North Sea. On the Dogger Bank young plaice grow rapidly and quickly attain good marketable size, becoming thick and of good flavour. Their offspring in the egg and larval stages drift passively with the currents and practically all are carried away from the Dogger, mostly north-eastwards to the extensive shallow sandy area off the Danish and German coasts. Here they find suitable food to develop to a small size and those which remain in the area continue growth only at a slow rate. Others move out into deeper water, and of these the plaice which find their way back to the Dogger Bank increase in size rapidly. A number caught off the Danish coast, where they are very numerous, were marked and liberated on the Dogger Bank. Here they increased in weight some six times as fast as others which were liberated where they were caught, in the coastal area.

Average weight of marked plaice when liberated in April	$2\frac{1}{2}$ oz.
Average weight in November after seven months in the coastal area	$4\frac{1}{5}$ oz.
Average weight in November after seven months on the Dogger Bank	15 oz.

Another instance where the animal population is controlled by physical conditions may be cited ((21) 1925). A patch of ground on the sea bottom of Bigbury Bay on the South Devon coast, consisting of silt and sand and forming a bank, was in the summer of 1922 densely populated with small flat fish. These were feeding on a small cockle-like shellfish, *Spisula elliptica*, which was present in immense numbers on the bottom. Practically all had been deposited there as 'brood' in the spring of the same year.

The number of these *Spisula* was considerably less next summer, many having died or been eaten by flat fish, crabs and starfish. In the spring of 1923 the larvae did not settle in any number on this suitable patch of ground, having been carried away by currents, or breeding having been restricted by other physical causes. Hence there was only a small 'brood' in the spring of 1923.

In 1924 the original 1922 *Spisula* were still fewer, and in the spring again little 'brood' had settled on the ground. From a fertile rich area abounding in flat fish in 1922, the ground had become poor in 1924.

All animal life entails loss of energy owing to the metabolic changes within the organism during its lifetime: the energy of light stored up in the algae during photosynthesis is thus dissipated by the animals which absorb them as food. Carbon dioxide, combined nitrogen and phosphorus are continuously being lost in the processes of respiration and excretion.

The rate of loss of the end products of basal metabolism is dependent upon the *temperature* in the case of cold-blooded marine animals whose body temperature is close to that of the water in which they are living. Very numerous experiments with a large variety of such organisms go to show that a rise of 1° C. increases the rate of metabolism by roughly 10 %, that is to say that the rate of oxygen intake and CO_2 output about doubles for a rise of 10° C. This increase in rate of metabolism per degree varies with different marine animals; the value 10 % is only a rough index of the magnitude.

Table VI. Rate of oxygen absorption with temperature (Vernon).

Rate of O_2 absorption	Temp. °C.							
	10°	12°	14°	16°	18°	20°	22°	24°
By *Beroe ovata*	0·40	0·58	0·78	1·00	1·23	1·47	1·75	2·04
By *Amphioxus lanceolatus*	0·58	0·72	0·86	1·00	1·14	1·28	1·42	1·56

In warm latitudes where animals have a considerably greater rate of metabolism than in cooler waters, they will require more food daily to repair the wastage due to their metabolic processes. It follows that the same amount of plant life will support more animals in higher latitudes than in the tropics.

Such an increase of rate with temperature is not limited to metabolism alone, it applies also to numerous other biological processes such as the rate of ciliary movement, of movement by amoebae, of enzyme action, of respiratory movements, of heart beat. The magnitude of the increase is usually of the same order as the increase in metabolism, within the range of temperature to which the animals are accustomed in their usual surroundings.

As an instance, Johansen and Krogh ((22) 1914) have determined the time taken for plaice eggs to hatch out after fertilisation and for the larvae to attain a length of 4·6 mm. at various temperatures.

Table VII.

Temperature ° C.	Days
4·1	23·0
6·1	18·1
8·0	13·3
10·1	10·3
12·0	8·3

Similar experimental results were obtained by Dannevig. In these cases the rate of development is a linear function of the temperature and increases by 20 % per 1° C. rise. In similar experiments with other fish the effect of temperature was not so great.

This relatively enormous effect of temperature on the physiological processes in marine animals is also well illustrated by an experiment made by Dr L. Hogben and communicated to the writer. The excised heart of a spider crab, *Maia squinado*, was bathed with a saline solution at varying temperatures, and the rate and amplitude of the beat recorded. At 4° C. the heart practically ceased to beat, and above 21° C. it appeared to be

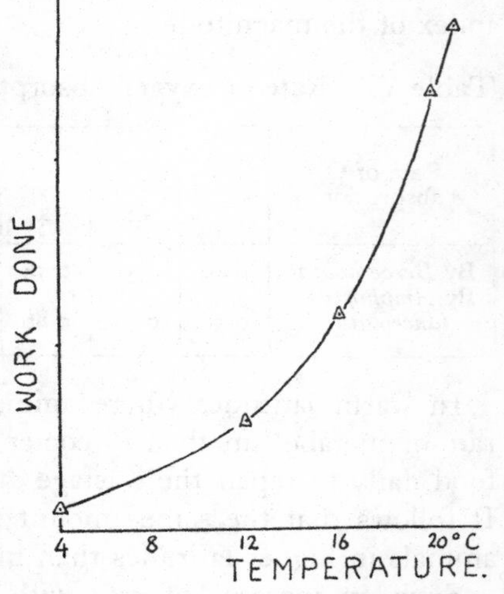

Fig. 2. Relation between work done in unit time by excised heart of *Maia squinado* and temperature. (From kymograph record made by Dr L. Hogben, July, 1924.)

injured. At temperatures between 4° and 21° C. *both* rate and amplitude increased with rise in temperature.

The work done by the heart muscle in unit time will be approximately proportional to the product of the recorded amplitude of the beat multiplied by the number of beats per unit time, in this case where it is working against the weight of the recording lever. This product is plotted against temperature in Fig. 2, and shows the magnitude of the temperature effect on the heart muscle.

These considerations indicate that, within the limits of temperature experienced by an animal in the sea, its 'rate of living' is increased in the order of 10 % for every degree centigrade rise in temperature; its requirements—oxygen and food—are increased in proportion.

Orton ((23) 1920) has shown that the breeding season of many marine animals lies within comparatively small limits of temperature. Variations in the temperature of the sea between one year and another may materially alter the duration and time of the breeding season.

That the fauna is exposed to a very different range of temperature from year to year is well shown in the following table, which refers to the bottom water, 22 miles south-west of Plymouth:

Table VIII.

Temperature of bottom water (° C.)	1921	1922	1923	1924
Over 15	Early Oct. to early Nov.	Never	Never	—
,, 14	Early Sept. to early Dec.	Mid. Sept. to mid. Oct.	Never	Never
,, 13	Early July to mid. Dec.	End Aug. to early Nov.	Mid. Sept. to mid. Oct.	Late Sept. to end Oct.
Below 11	—	End Jan. to end June	Mid. Dec. 1922, to end June	Early Dec. 1923, to early July
,, 10	—	Early March to mid. May	Mid. Jan. to end April	Early Dec. 1923, to end May
,, 9	--	Never	Never	End Jan. to early May

A similar fauna may be expected to inhabit regions having the same water temperature with an equal range. In this connection it is interesting to read in the late Sir John Murray's summary of the *Challenger* Expedition: "The general similarity between the animals from dredgings and trawlings in high southern latitudes and those from high northern latitudes, was the subject of frequent discussion among the naturalists during the whole of the southern cruise." Mayer ((24) 1914) also remarks that: "Probably no single factor is a more extensive barrier to extended geographical range of marine animals than is that of temperature. This fact has long been recognised and there are many striking examples of its general truth; for example, Professor Hjort concludes that the southern limit of the northern boreal species of fishes from the sea bottom everywhere coincides with the 10° C. isotherm at a depth of 100 metres. That temperature should be so important a barrier to universal distribution is the more remarkable in view of the large number of observations showing how readily animals may become artificially adapted to survive in unaccustomed temperatures." He quotes an experiment by Dallinger (1887) who, by gradually raising the temperature over several years, acclimatised a colony of Flagellata to 70° C., which normally were killed at 23° C.

It is a notable fact that many marine animals live, particularly in tropic seas, at a temperature very little below that which causes their death.

In many cases the rate of metabolism, or of other vital reactions, is not a linear function of the temperature, but the increase in rate per rise of 1° C. itself changes as the temperature rises. It is customary to express the increase in rate with temperature by the term Q_{10}, which denotes $1 + \frac{1}{10}$ of the percentage increase in rate for a rise of 1° C. As stated above, the value of Q_{10} is frequently not constant, but changes with increasing temperature. An instance of this is the case of simple move-

Table IX.

° C.	Interpolated temperature coefficients (Q_{10}) for the rate of movement of	
	Amoebae	Cilia
0–5	16·9	3·52
5–10	3·19	3·00
10–15	2·31	2·37
15–20	2·04	2·25

ments of protoplasm in amoebae and the movement of cilia. Here the fall in viscosity of the protoplasm per degree rise in temperature is much more rapid at lower than at higher temperatures, and the value of Q_{10} for the increase in rate decreases as the temperature rises ((25) 1924).

Crozier has advanced an argument that the rate of many biological reactions is the rate of the slowest irreversible chemical reaction in the series of chemical reactions which take place.

The effect of temperature upon an irreversible chemical reaction as stated by Arrhenius is given by the equation

$$\frac{K_2}{K_1} = \epsilon^{\frac{\mu}{2}\left(\frac{1}{T_1} - \frac{1}{T_2}\right)},$$

where K_1 is the velocity at $T_1°$ Abs. and K_2 the velocity at $T_2°$ Abs., μ being a constant for the particular reaction, and $\epsilon = 2{\cdot}718....$

From collected data of the effect of temperature on the rate of various biological processes, Crozier concludes that this relation is generally followed through a part at least of the range of temperature investigated, breaks occurring when the value of μ alters owing to a change over of the slowest chemical reaction. Since the value of μ is not only an index of the particular irreversible chemical reaction, but also from theoretical grounds probably equals the heat of reaction in calories, Crozier has deduced the nature of the slowest chemical reaction (as the oxidation of carbohydrate or of fat, etc.) which limits the rate of the biological process. Two objections have been raised concerning the validity of these conclusions: the experimental evidence is insufficient for the conclusions respecting the constancy and value of μ which have been drawn; the velocity of the assumed limiting chemical reaction is itself influenced by the velocity of other reactions in the series which affect the rate of supply of the reacting substances and the concentration of its end products.

In addition to these two physical factors—currents and temperature—a third physical factor, light, affects the fauna both indirectly and directly. Fluctuations in the seasonal variation of light undoubtedly exert an effect upon the fauna in some instances through the supply of phytoplankton food.

A series of events in higher latitudes, where there is a considerable seasonal variation in temperature and light, possibly influences the survival of young fish, at the stage when they have just absorbed the yolk sac, and perhaps of other animals, which are dependent upon diatoms for their nutrition for some days or weeks. In such regions the time of onset of the burst of diatom growth varies with the rather sudden increase in daily sunshine which occurs about February or March. The time at which the particular fish become dependent upon diatoms is determined rather by the temperature of the water, which does not vary year by year with the hours of sunshine, but is the cumulative result of past meteorological events and inflowing drifts. Hence, as pointed out by Hjort, it may so happen that in some years the young fish arrive at a time when their requisite food is plentiful, in other years before or after the glut.

Another case of fluctuations corresponding with the seasonal variation in light may be quoted.

A regular fishery for mackerel takes place in the vicinity of Land's End, and by far the largest number are landed during May. These mackerel feed on animal plankton (copepods) and it was noticed that during May of those years when copepods were most numerous, mackerel appeared to be present in greater numbers. Investigations had indicated that the movement of the water masses in the area is usually comparatively slow, hence the copepods found there in May were for the most part passing through their young stages during March and April in the same area. They feed on phytoplankton, diatoms and peridinians, and it has already been shown that the onset of growth of these plants corresponds with the amount of sunshine. Thus in a year with much sunshine in February and March the phytoplankton organisms will be more numerous in March and April than in a year with delayed sunshine. That is to say, during the years with most sunshine in February and March the copepods occurring in May will have had more food. Allen ((26) 1909) has obtained data of the number of mackerel caught during May by three steam drifters fishing on this area throughout the month. On plotting the average number of mackerel caught in May each year against the hours of sunshine occurring in February and March, a remarkably close correlation was found (Fig. 3).

Quite apart from the effect of illumination in determining the production of the plants which provide the ultimate food source of all animal life, light exerts a marked influence upon many marine animals.

It has long been known that some free-swimming plankton animals in a vessel of water will congregate where the illumination is greatest, while other species congregate where it is least. The migration of many species to the surface of the sea during the night and their withdrawal during daylight to deeper layers has been repeatedly observed. The recent quantitative observations by Russell have shown that numerous species of free-swimming animal plankton organisms and young fishes are distributed in a more or less regular manner dependent upon the strength of illumination at different depths, the greatest number of a particular species occurring at a depth which moves upward as the light decreases in the evening and downward as the light

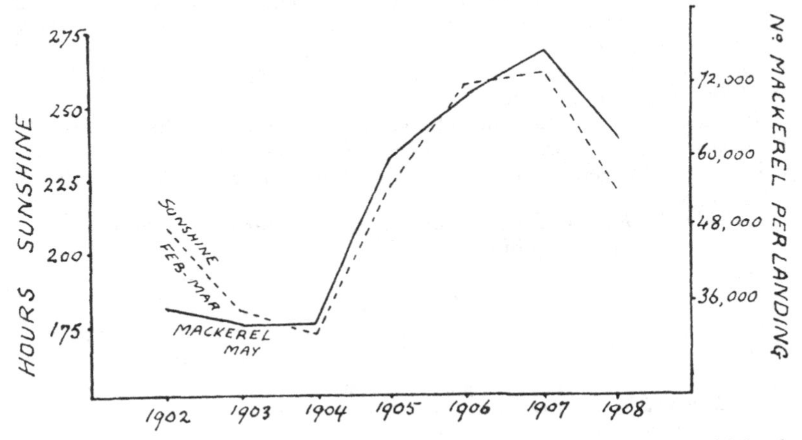

Fig. 3. Dotted curve shows the hours of sunshine during February and March in each year from 1902 to 1908 (average of values recorded at Plymouth, Falmouth and Scilly). Continuous curve shows the average number of mackerel per vessel landed during May by three steam drifters.

increases during the early hours of the morning. For some species it has been observed that the depth at which the majority occur at midday at midsummer is greater than at midday a month before and after midsummer, when the light penetrates less deeply owing to the greater obliquity of the rays and lesser total illumination. Other species have shown no such influence of the illumination upon their distribution with depth ((27) 1927).

Passing next to the influence of salts occurring in sea water upon marine animals, a most striking feature of sea water is the constancy of the relative proportions in which its constituents occur. It is improbable that a lack of any salts other than nitrates and phosphates affects the population in nature. Sensible differences in the total salt content or salinity do however occur, and the distribution of many marine animals has been noticed to lie within fairly well-defined limits of total salinity. The blood of elasmobranch fishes and marine invertebrates is almost of the same osmotic pressure as that of the water in which they live (28), and rapidly adjusts itself to a change in salinity.

Macallum ((29) 1926) mentions that the freezing-point of the blood of the elasmobranch fish *Acanthias vulgaris* was found to be $-2{\cdot}035^\circ$ C. while that of the water in which it lived ranged between -2° and $-2{\cdot}3^\circ$ C. However, of this depression, amounting in all to $2{\cdot}035^\circ$, only $1{\cdot}075^\circ$ was due to the salts in solution in the blood, the remaining $0{\cdot}060^\circ$ being largely due to urea, which occurs in considerable concentration in the blood of elasmobranchs.

It must be noted that the relative concentrations of the salts is not necessarily the same in blood and body fluids as in sea water, although it is similar, as seen from the following ratios quoted by Macallum:

Table X. Ratios of ions occurring in sea water and in the blood and body fluids of various marine animals.

	Na	K	Ca	Mg	SO_3	Cl
Ocean water	100	3·61	3·91	12·10	20·9	180·9
Limulus	100	5·62	4·06	11·20	13·39	186·9
Aurelia flavidula	100	5·18	4·13	11·43	13·18	185·5
Homarus americanus	100	3·73	4·85	1·72	6·67	171·2
Acanthias vulgaris	100	4·61	2·71	2·46	—	165·7
Gadus callarias	100	9·50	3·93	1·41	—	149·7
Pollachius virens	100	4·33	3·10	1·46	—	137·8

Many marine animals including fish can live and flourish in water having a wide range of salinity, and many fish move into estuarine and even fresh water during their breeding season. The recent Cambridge zoological survey of the Suez Canal has shown that several Mediterranean and Indian Ocean species have acclimatised themselves and form a flourishing community in the very saline water of the Great Bitter Lake.

In nature, water of lower salinity than the neighbouring water masses is often subject to a greater annual range of temperature (p. 142). Since salinity distribution tends to march with temperature distribution in the sea, and temperature is known to have an all-important influence on marine organisms, their distribution in the open sea is more likely to be influenced by temperature than by salinity, which may however act towards the animal as an index of the seasonal range of temperature the water is likely to undergo. The differences in salinity occurring in the open oceans are appreciable but not great; it is only when definitely estuarine or definitely arctic conditions are met with that big changes occur.

It is remarkable that certain marine animals have the power to absorb salts occurring in infinitely small quantities in sea water—such as strontium, vanadium, nickel, cobalt and very frequently copper, which latter plays a necessary part in the life processes of many invertebrates. It is for this reason that as complete a list as possible of these metals occurring in traces is given on p. 51, although, except in the case of iron and copper, the 'reason' why they are absorbed is quite obscure.

The part played by the traces of filter-passing organic matter, occurring in sea water, upon the physiological processes of marine animals in nature, is by no means clear. Pütter ((30) 1909) has advanced a theory that they absorb much of their nutriment in the form of dissolved organic matter, but the theory has not been generally accepted on the ground that much subsequent experimental evidence has failed to show any such absorption, and that there is plant life sufficient to support them. The *capability* of higher marine organisms to obtain nutriment by absorbing dissolved organic matter themselves is not definitely proved or disproved ((31) 1925). That marine bacteria and possibly many protozoa utilise these dissolved substances and keep their concentration low in the open sea is probable, if not certain.

Resulting from an investigation of the quantity of organic matter in solution in the water of Oslo Fiord, Gran makes the interesting suggestion that small phytoplankton organisms lose much of the sugars, etc., produced in photosynthesis, by diffusion out into the surrounding water, and that this organic matter in solution is utilised by colourless dinoflagellates and infusorians which occur in great numbers after the outbursts of diatom growth ((39) 1926). It is improbable, however, that the diatoms are permeable before they die or form spores.

The question of the utilisation of dissolved organic matter occurring in the sea by organisms cannot be lightly dismissed. In inshore waters, at all events, there is several times as much organic matter in solution as is present in the living organisms in that water.

An investigation by Peters ((32) 1921) has indicated that the ciliate *Colpidium* in a culture free from bacteria absorbed nutritive substances in solution, and that it can utilise either amino acids or ammonium salts in building up its own proteins.

In this connection it is also of interest to consider the possibility of dissolved organic matter being utilised directly by marine algae, although owing to the very low concentration in the sea it is not suggested that this means of nourishment plays a material part in nature, even if it takes place at all.

A green unicellular alga occurring in the soil—*Scenedesmus*—has been shown by Roach ((33) 1926) to grow more luxuriantly in liquid media to which certain soluble organic compounds such as glucose and other sugars have been added, than in liquid media without such addition. Further this alga actually grows in complete darkness in a nutritive salt solution enriched with glucose and does not lose its chlorophyll for several months.

Miquel found that some bottom-living diatoms would grow better in sea water enriched with nutrient salts to which some soluble organic matter, such as an infusion of algae, had been added than without this addition, but Allen and Nelson ((19) 1910) did not find that such an addition improved the growth of plankton diatoms. However they do note instances in which weak organic infusions added to sea water without added salts greatly improved the growth of plankton diatoms. In this case the diatoms were not "necessarily feeding on dissolved organic material, as some necessary saline nutritive materials could have dissolved out from the weed or fish" from which the infusion was made.

The general fertility of the narrow zone extending from half tide mark to a few feet below low tide mark around coasts is probably due to the abundant food for protozoa, bacteria and other lowly organisms, provided by the detritus from dead seaweed and land drainage, together with the soluble organic matter leached out of them. The minute traces of organic matter in sea water are known to be greater closer inshore (p. 53). The protozoa, etc., so nourished provide in themselves carbohydrate and protein food for more highly developed organisms. Hence as far as the general fertility of the area is concerned, the capacity or incapacity of highly developed marine animals, fishes, etc., themselves to utilise the traces of organic matter in solution is apparently a question of little economic importance. The work of building these substances up into carbohydrate and protein can be carried out by proxy.

Blevgard ((34) 1914), from observations mostly on the *Zostera* detritus of the shallow Danish fiords, reaches the conclusion that organic detritus forms the principal food of the great majority of animals on the sea floor. This area is an extreme case, being shallow and extensive with the bottom covered with eel grass—*Zostera*—unlike the narrow fringe of seaweeds around most coasts. The more recent observations of Hunt ((35) 1925) suggest that in deeper water the protozoa and bacteria in the detritus form the chief and more readily assimilated food of the larger animals on the sea floor, and that these are not directly nourished by the less digestible pentosans and almost negligible traces of digestible proteins found in the detritus itself ((36) 1914).

As the shore is approached the number of bacteria in the upper layers increase. Bertel ((37) 1912), working in the Mediterranean, has observed that in the open sea away from the coast the bacteria increase in number with depth. At the surface they are mostly killed by strong sunlight, to be replaced by others during the night. He quotes an interesting observation by Di Giaxa that most pathogenic bacteria are capable of living and multiplying in sterilised sea water, but in water from Naples harbour containing a rather rich natural bacterial flora they rapidly succumb in the struggle for existence.

The products of organic decay from rivers and towns may however affect the fauna of estuarine and inshore waters. When typically open sea animals are kept in small aquaria, in clean water from the open sea and at constant or nearly constant temperature, and fed upon natural food as nearly as possible, they are extremely susceptible to the least decay of organic matter taking place in the water. Garstang reared blennies from the egg in the Plymouth laboratory; it was only by the most meticulous care in taking out of the vessel of water any particle of dead organic matter, excreta, etc., which was liable to putrefaction, and by frequent changes of water, that the larvae could be reared. The writer has kept a typically open sea copepod—*Calanus finmarchicus*—in small vessels of water and noted that their length of survival is little affected by quite wide variations in salinity or carbon dioxide tension, but that the development of putrefactive bacteria in the water rapidly kills them. Further, if sea water is polluted, as by the addition of peptone or a dead worm, and after putrefaction has died down the water is filtered through a Berkfeldt candle, it is still toxic and remains so for some time, even after subsequent filtrations, although not nearly so toxic as water in which putrefaction is actually taking place ((38) 1925). It has long been customary to filter the water of small aquaria through animal charcoal or to treat it with hydrogen peroxide, in order to keep the water 'fresh' by absorbing or oxidising the products of decay.

The animals which can most readily be kept in aquaria are, for the most part, those which live in shallow depths along the coast and in estuaries.

BIBLIOGRAPHY

(1) LOHMANN, H. "Die Bevölkerung des Ozeans mit Plankton u.s.w." *Archiv f. Biontologie*, **4**, Heft 3. Berlin, 1920.

(2) GRAN, H. "Pelagic Plant Life" in *Depths of the Ocean*. Murray and Hjort. London, 1912.

(3) ATKINS, W. R. G. "The Hydrogen-Ion Concentration of Sea Water. I." *Journ. Mar. Biol. Assoc.* **12**, 717. 1922.

—— "The Phosphate Content of Fresh and Salt Waters." *Journ. Mar. Biol. Assoc.* **13**, 119. 1923.

—— "A Quantitative Consideration of some Factors concerned in Plant Growth in Water." *Journ. du Cons. Int. pour l'Expl. de la Mer*, **1**, 197–226. 1926.

MOORE, B., PRIDEAUX, E. and HERDMAN, G. "Studies of certain photosynthetic phenomena in sea water." *Proc. and Trans. Liverpool Biol. Soc.* **29**, 233–64. 1915.

(4) BAUER, E. "Ueber zwei denitrificirende Bakterien aus der Ostsee." *Wiss. Meeresuntersuchungen*, **6**. Kiel, 1901.

FEITEL, R. "Beiträge zur Kenntniss denitrificierender Meeresbakterien." *Wiss. Meeresuntersuchungen*, **7**. Kiel, 1903.

See also DREW, C. H., "On the Precipitation of Calcium Carbonate in the Sea by Marine Bacteria, and on the Action of Denitrifying Bacteria in Tropical and Temperate Seas." *Journ. Mar. Biol. Assoc.* **9**, 479–524. 1913.

(5) DOMOGALLA. "The forms of Nitrogen found in certain Lake Waters." *Journ. Biol. Chem.* **63**, 269–85. 1925.

(6) SCHREINER and SKINNER. *U.S. Dept. Agri. Bureau of Soils Bull.* **87.** 1912.
(7) MOORE, B. *Biochemistry.* London, 1921.
(8) BRISTOL, M. and PAGE, H. "A Critical Enquiry into the Alleged Fixation of Nitrogen by Green Algae." *Ann. Applied Biology*, **10**, 378. 1923.
(9) REINKE, E. "Zur Kenntnis der Lebensbedingungen von Azotobacter." *Ber. Deut. Bot. Gesells.* **22**, 95. 1904.
(10) KEUTNER, J. "Über das Vorkommen u. die Verbreitung Stickstoffbindender Bacterien im Meere." *Wiss. Meeresuntersuchungen*, **8**, 27. Kiel, 1905.
(11) KEDING, M. "Weitere Untersuchungen über Stickstoffbindende Bacterien." *Wiss. Meeresuntersuchungen*, **9**, 275. Kiel, 1906.
(12) BRANDT, K. "Über den Stoffwechsel im Meere." *Wiss. Meereskunde*, **18**. Kiel, 1916–1920.
(13) BRIGGS, G. E. "Characteristics of Subnormal Photosynthetic Activity resulting from Deficiency of Nutrient Salts." *Proc. Roy. Soc.* **94**, B, 20–35. 1922.
(14) KNEIP and MINDER. "Über den Einfluss verschiedenfarbigen Lichtes auf die Kohlensäureassimilation." *Zeit. f. Bot.* **1**, 619–50. 1909.
WURMSER, R. "L'action des radiations des différentes longueurs d'onde." *Compt. Rend.* **170**, 1610–12. 1920.
(15) KLUGH, A. B. "The Effect of Light of Different Wave Lengths on the Rate of Reproduction of *Volvox aureus* and *Closterium acerosum*." *New Phytologist*, **24**, 186–90. 1925.
(16) PLAETZER, H. "Untersuchungen über die Assimilation und Atmung von Wasserpflanzen." *Verhand. phys.-med. Ges. Würzburg*, **45**, 31–102. 1917.
KANITZ, A. *Temperatur und Lebensvorgänge.* Berlin, 1915.
(17) MATTHAEI, G. "The Effect of Temperature on Carbon Dioxide Assimilation." *Proc. Roy. Soc.* **72**, 355. 1904.
OSTERHOUT and HAAS. "The Temperature Coefficient of Photosynthesis" (*Ulva*). *Journ. Gen. Physiol.* **1**, 295. 1918.
BLACKMAN, F. F. "Limiting and Optima Factors." *Ann. Bot.* **19**, 281. 1905.
(18) KNEIP, H. "Assimilation u. Atmung der Meeresalgen." *Internat. Rev. Hydrobiol. u. Hydrog.* **7** (1), 1–37. 1914.
(19) ALLEN and NELSON. "On the Artificial Culture of Marine Plankton Organisms." *Quart. Journ. Micr. Sci.* **55**, 361–431. 1910.
(20) ALLEN, E. J. "On the Culture of the Plankton Diatom *Thalassiosira gravida*, Cleve, in Artificial Sea Water." *Journn. Mar. Biol. Assoc.* **10**, 417–39. 1914.
(21) FORD, E. "On the Growth of Lamellibranchs in Relation to the Food Supply of Fishes." *Journ. Mar. Biol. Assoc.* **13**, 531–59. 1925.

(22) JOHANSEN, A. G. and KROGH, A. "The Influence of Temperature and certain other factors upon the Rate of Development of the Eggs of Fishes." *Publ. de Circonstance*, No. 68. 1914.

(23) ORTON, J. H. "Sea Temperature, Breeding and Distribution of Marine Animals." *Journ. Mar. Biol. Assoc.* **12**, 339. 1920.

(24) MAYER, A. G. "Effect of Temperature on Tropical Animals." *Publ.* 183, *Carnegie Inst. Washington*, 1–24. 1914.

(25) PANTIN, C. F. A. "On the Physiology of Amoeboid Movement." *Brit. Journ. Exp. Biol.* **1**, 519. 1924.

—— "Temperature and the Viscosity of Protoplasm." *Journ. Mar. Biol. Assoc.* **13**, 331. 1924.

(26) ALLEN, E. J. "Mackerel and Sunshine." *Journ. Mar. Biol. Assoc.* **8**, 394–406. 1909.

(27) RUSSELL, F. S. "Vertical Distribution of Plankton in the Sea." *Biological Reviews*, **2**, 213–262. 1927.

(28) DAKIN, W. J. "Aquatic Animals and their Environment. The Constitution of the External Medium and its Effect upon the Blood." *Internat. Rev. Hydrobiol. u. Hydrog.* **5**, 53. 1912.

—— "Variation in the Osmotic Concentration of the Blood and Coelomic Fluids of Aquatic Animals caused by Changes in the External Medium." *Biochem. Journ.* **3**, 473. 1908.

BOTTAZZI, F. "Osmotische Druck der Flussigkeiten der Einzelligen pflanzlichen und tierschen Organismen." *Ergebnisse der Physiol.* **7**, 246–62.

(29) MACALLUM, A. B. "Palaeochemistry of Body Fluids and Tissues." *Physiological Reviews*, **6**, 316. 1926.

(30) PÜTTER, A. *Ernährung der Wasserthiere.* Jena, 1909.

—— "Der Ernährung der Wasserthiere durch gelöste organische Verbindungen." *Arch. für die ges. Physiol.* **137**, 595. 1911.

(31) DAKIN and DAKIN. "The Oxygen Requirements of Certain Animals and its bearing on the Food Supply." *Brit. Journ. Exp. Biol.* **2**. 1925.

KRIKENSKY, J. "Ueber die nutritive Bedeutung der im Wasser aufgelösten organischen Substanzen für die Wassertiere." *Archiv. Hydrobiol.* **16**, 169. 1925.

(32) PETERS, R. A. "The Substances needed for the Growth of a Pure Culture of Colpidium Colpoda." *Journ. of Physiol.* **55**, 1. 1921.

(33) ROACH, B. M. B. "On the Relation of Certain Soil Algae to some Soluble Carbon Compounds." *Ann. Bot.* **40**, No. CLVII, 149. 1926.

(34) BLEVGARD, H. "Food and Conditions of Nourishment among the Communities of Invertebrate Animals found on or in the Sea Bottom in Danish Waters." *Rep. Danish Biol. Station*, **22**. 1914.

(35) HUNT, O. D. "Food of the Bottom Fauna of the Plymouth Fishing Grounds." *Journ. Mar. Biol. Assoc.* **13**, 560–99. 1925.

(36) BOYSEN JENSEN. "Studies Concerning the Organic Matter of the Sea Bottom." *Rep. Danish Biol. Station*, **22**. 1914.

(37) BERTEL, R. "Sur la distribution quantitative des bactéries planctoniques des côtes de Monaco." *Bull. Inst. Océanogr. Monaco*, No. 224. 1912.

(38) HARVEY, H. W. "Oxidation in Sea Water." *Journ. Mar. Biol. Assoc.* **13**, 953–69. 1925.

(39) GRAN, H. H. and RUUD, BIRGTHE. "Untersuchungen über die gelösten organischen Stoffe." *Avhand. utgiff av Det Norske V.-Akad. i Oslo.* I. Matem.-Naturvid. Klasse, No. 6. 1926.

GAARDER, T. and GRAN, H. "Production of Plankton in Oslo Fiord." *Rapp. et Procès-Verbaux du Conseil Perm. Internat. pour l'Exploration de la Mer*, **42**. Copenhagen, 1927.

Chapter II

CHEMISTRY OF SEA WATER

SALINE CONSTITUENTS

Numerous analyses of the salts in sea water from various localities have been carried out. Clarke ((1) 1924) has tabulated 24 such analyses, chosen from the most reliable sources. A strikingly close agreement is shown among those of water from the open oceans. The concentration of the mixture of salts varies considerably from about 3·8 % in the water of the Sargasso Sea to a low value when diluted with fresh water from melting ice in the polar regions, or from rivers in the neighbourhood of land, but the proportion of the constituents making up the total salt content

Table XI. Analyses of salts in ocean water.

	Percentages		
	A	*B*	*C*
Cl	55·29	55·18	55·25
Br	0·19	0·18	—
SO_4	7·69	7·91	7·56
CO_3	0·21	0·21	0·37
Na	30·59	30·26	30·76
K	1·11	1·11	1·14
Ca	1·20	1·24	1·22
Mg	3·72	3·90	3·70

A. Mean of 77 analyses from many localities collected by *Challenger* Expedition. W. Dittmar, '*Challenger*' *Rep.* **1**, 203. 1884.
B. Mean of 22 samples collected between Cape of Good Hope and England. Makin, *Chem. News*, **77**, 155, 171. 1898.
C. Mean of 5 samples collected near Beaufort, N. Carolina. Wheeler, *Journ. Am. Chem. Soc.* **32**, 646. 1910.

of waters in the great oceans, Atlantic, Pacific and Indian, is practically constant.

From a number of the most careful determinations the following relation has been arrived at by Knudsen:

$$S\ ‰ = 0{\cdot}030 + 1{\cdot}8050\ Cl,$$

where S ‰ very closely approximates to the total weight of salts in grams per 1000 grams of sea water, generally known as the *salinity* of the water, and Cl is the weight of chlorine in grams per 1000 grams of sea water (chlorine plus the chlorine equivalent of the bromine and iodine in the sea water is actually measured).

The relation between total salts and chloride given above is an approximation; for the water of the Mediterranean a slightly different proportion exists. It should be mentioned that the accuracy of the above formula connecting salinity with chlorine content has been recently questioned by Giral ("Quelques observations sur l'emploi de l'eau normale en océanographie," *Publications de Circonstance*, No. 90) and also the accuracy to be obtained by the experimental procedure in ordinary use. However, the consistency of the results obtained by several observers with waters from the Atlantic and Pacific indicates that the experimental results on which this criticism is based are considerably outside the degree of accuracy ordinarily attained.

In order to arrive at the distribution of density of the water for some hydrographical purposes, where the movements of the water masses in the ocean are being investigated, it is required to estimate the salinity of samples of sea water to the greatest degree of accuracy.

A method has been developed for the titration of sea water with silver nitrate solution whereby the chlorine content can be estimated to within 0·01 parts per 1000 of water and the salinity to 0·02 parts or less per 1000. A full description of the method is given by Oxner and Knudsen ((2) 1920). It depends upon comparison of the volume of silver nitrate solution required to give a faint red tint to 15 (or 10) c.c. of sea water, to which has been added 6 drops of 8% potassium chromate solution, with the volume required by an equal amount of a sea water of definitely known chlorine content. 'Normal water' for this purpose is supplied by the Laboratoire Hydrographique of the Conseil Permanent International pour l'Exploration de la Mer, Copenhagen, and generally contains 19·38‰ Cl, the exact value being marked on each tube. By this arrangement workers in the different countries have the same basis for comparison. The calculation is carried out with the aid of Knudsen's *Hydrographical Tables* (3).

For many purposes, to arrive at the salinity of a sample of sea water, it is sufficient to titrate 10 c.c. with a solution containing 27·25 grm. of silver nitrate per litre from an ordinary burette.

The volume of c.c. of the silver nitrate required will roughly equal the salinity of the sample. To obtain a more accurate result an equal volume of 'normal sea water' of known salinity is titrated at the same temperature (about 15° C.), and from the two amounts of silver nitrate solution required, the salinity of the unknown is calculated, assuming that the salinities are proportionate to the volumes of silver nitrate solution used. Actually they are not quite in the same proportion since 10 c.c. of a more dilute sea water than the 'normal' water will not weigh so much. To allow for this it is necessary to apply a small correction given in the following table.

Table XII.

Salinity, S°/oo, found	Correction to be applied	Salinity, S°/oo, found	Correction to be applied
40	−·15	22	+·22
38	−·08	20	+·23
36	−·03	18	+·23
34	+·03	16	+·23
32	+·07	14	+·20
30	+·11	12	+·19
28	+·15	10	+·16
26	+·17	8	+·15
24	+·20	6	

It is necessary during the addition of silver nitrate to stir vigorously, otherwise the flocks of silver chloride tend to hold salt water enmeshed within them, and it is essential to keep the burette and pipette scrupulously clean and free from grease by washing them with potassium bichromate dissolved in concentrated sulphuric acid, otherwise the sea water and silver nitrate solution gather unevenly in tears on the side of the pipette and burette respectively. The sea water and silver nitrate solution wetting the sides of the pipette and burette should be allowed approximately the same time to drain down in each titration.

The total salt content of the water may also be assessed from a determination of the density or specific gravity. Unfortunately the determination of density by ordinary and rapid means is insufficiently accurate for many purposes. The ordinary immersion hydrometer is subject to a variable error when in use, owing to the pull downwards at the meniscus decreasing if the surface of the liquid is in the least contaminated. A total immersion hydrometer gets over this error; with such there is a liability for bubbles of gas to form on the apparatus, if the sample of sea water is supersaturated at the temperature at which the determination is being carried out. This source of error can be obviated by working at a higher atmospheric pressure, by previously heating or by sucking some of the dissolved gas out of the water before making the determination. To obtain results with an accuracy of ± 0·01 S ‰ it is however necessary to know the temperature at which the density is measured to within 0·01° C., unless the hydrometer has a similar coefficient of expansion to the water. A rapid and accurate method is much needed. Numerous densimetric, electrical conductivity and refractive index methods have been devised, particularly with a view to carrying out the determination on board ship, but so far the titration method is in most general use.

The United States Coast Guard Service have developed the electrical conductivity method of measuring salinity and use it on board their vessels engaged on the International Ice Patrol. Owing to the various devices necessary to obtain the required accuracy the apparatus is expensive, and owing to the necessity for the sample of water being allowed to attain the exact temperature of a bath in which the resistance cell is immersed, each determination takes about ten minutes. (*Bulletin No.* 12, *U.S. Coast Guard*, p. 136, Washington, 1924.)

An instrument, based on Lord Rayleigh's Interferometer method of measuring the refractive index of gases, has been manufactured by the firm of Zeiss for the measurement of the refractive index of liquids and it is stated that rapid and accurate determinations can be made of the index of sea water, to which the density and salinity are directly related. (Pape, C., "Über Interferometers zur Bestimmung des Salzgehalts von Meerwasser," *Ann. Hydrog. und Maritimen Meteorologie*, p. 291, 1925; Schumacher, A., "Die Deutsche Atlantische Expedition 'Meteor'," *Zeitschrift der Gesellschaft für Erdkunde zu Berlin*, Jahr. 1926, No. 1, p. 47, and 1926, No. 5/6, p. 251.)

The relation between salinity (in parts per 1000 by weight), specific gravity and the variation in density with temperature, is given in Knudsen's *Tables*.

When water is subjected to the great pressure which occurs in the depths of the sea it contracts. The relation between the specific gravity when brought to the surface and when at various depths in the ocean has been tabulated by Ekman ((4) 1908, 1910).

Phosphates in Solution

The phosphates in the water off Plymouth were estimated by Matthews by means of the colorimetric method of Pouget and Chouchak ((5) 1916, 1917), and in North Sea waters by Raben, gravimetrically ((6) 1916–1920).

Atkins ((7) 1923–1926) has carried out very numerous analyses by the rapid colorimetric method of Denigès, obtaining values comparable to those of Matthews but lower than Raben's. A seasonal variation in the water of the English Channel was found, confirming the earlier results of Matthews, the water containing 30 to 40 mg. P_2O_5 per cubic metre in winter and being almost entirely depleted of phosphate in the summer.

The depletion occurs first in the upper layers, at a time in the beginning of the year when sunshine begins to exceed about three hours per day. The time varies from year to year; thus in 1924 it occurred a month to six weeks earlier than in 1923, the early months of which were less sunny. Towards the end of summer phosphate begins to be re-formed from dead organisms in the bottom water faster than it can be utilised by phytoplankton in the deep water, and by the end of October fairly complete vertical mixing of the surface with the bottom water has usually taken place. During each winter, 1923, 1924 and 1925, roughly the same amount of phosphate occurred in the water of the English Channel at the position investigated. This was almost completely used up in the upper layers each summer, the phosphate in the bottom water being reduced to about one-third of its winter value. A comparison of one year with another demonstrates that the main difference lies in the dates in early spring when the phytoplankton uses up phosphate at a rate greater than it is being re-formed from dead organisms, rather than in different amounts of phosphate present at the beginning of each year.

The annual variation in the amount of dissolved phosphate in the surface and bottom water at this position in the English Channel is shown in Fig. 62, p. 179. The effect of the amount of sunshine in March and April during different years on the dates at which the phosphate begins to be used up is seen on comparing these figures with Fig. 65 on p. 187.

The deep water of the open ocean, beyond the range of sufficient illumination for phytoplankton growth, contains a great store of phosphate, accumulated from the decay of dead organisms ((7a) 1926).

The surface water of the Mediterranean and of the Atlantic between England and South America was found to be almost entirely stripped of phosphate during the sunny months of the

Table XIII. Phosphate content of Atlantic water.

The samples were taken at lat. 37° 44′ N., long. 13° 21′ W. on October 12, 1925, save that at 3000 metres, which was taken on October 16 at lat. 29° 59′ N., long. 15° 03′ W.

Depth in metres	T. ° C.	*p*H	In mg. per cu.m.
			P_2O_5
0	21·10	8·35	0
10	21·10	—	—
20	21·00	—	—
30	21·00	—	—
40	21·00	—	—
50	20·01	8·35	0
75	17·31	8·31	5
100	15·10	8·18	8
150	15·06	8·16	10
200	13·86	8·11	22
300	12·25	8·12	44
500	10·94	8·00	50
1000	9·55	8·03	74
2000	4·81	7·94	78
3000	3·10	7·87	88

year. In the temperate zone in winter, and even in summer where the Atlantic current runs over the submarine ridge between Scotland and Iceland, the surface water is relatively rich in phosphate. Values found by Atkins in this area are given together with the nitrate content of the water on p. 182.

In the tropical regions of the South Atlantic the water below 300 metres contains from 200–300 mg. of P_2O_5 per cubic metre, falling to a vanishingly small amount at the surface, while in the Antarctic about 200 mg. per cubic metre occurs throughout (see p. 171).

In the Clyde sea area Marshall and Orr ((44) 1927) found that an increase in the number of phytoplankton organisms corresponded with a decrease in the amount of phosphate.

Atkins' Method of estimating Phosphates in Solution in Sea Water

Two c.c. of solution *A* and one drop of solution *B* are added to 100 c.c. of sea water in a white glass graduated cylinder. The same quantities are added to a solution of KH_2PO_4 in distilled water (0·050 mg. P_2O_5 per litre or other convenient standard) in a similar cylinder with a tap near the bottom. The lengths of the columns of the two liquids are then adjusted to give an equal tint of blue; from the length of the column and the strength of the standard the concentration of phosphate in the sample of sea water can be calculated.

Solution A. 100 c.c. of 10% ammonium molybdate in 300 c.c. of 50% (by volume) sulphuric acid. The reagent should be stored in the dark to minimise the production of a blue tint. With less sulphuric acid the reagent gives a blue tint with stannous chloride.

Solution B. This must be freshly prepared before carrying out the estimation by dissolving approx. 0·1 gm. tin in 2 c.c. concentrated HCl with one drop of 3–4% copper sulphate solution and making up to 10 c.c. On adding too much of this reagent a yellow tint tends to develop in the sea water, though not in the standard KH_2PO_4 solution, but by adding only the minimum amount of solution *B* this trouble is obviated—usually one drop to each 100 c.c. of sea water is sufficient.

Nitrates in Solution

Nitrates in very small amount have long been known to occur and their importance has been recognised for plant growth. The only method of determination until 1925 had been to reduce them by means of nascent hydrogen to ammonia, usually by adding aluminium amalgam.

The sea water, preserved against putrefaction of its contained organisms by the addition of mercuric chloride, was distilled with magnesium oxide in order to drive off all ammonia occurring as ammonium salts. The distillation was continued after digestion with amalgam, by which nitrate and nitrite had been converted into ammonia ((35) 1915). This led to results giving too high a concentration owing probably to the formation of ammonia from nitrogenous organic matter dissolved in the sea water. The formation of ammonia from impurities in the reagents and from nitrogen occluded in the amalgam necessitates the subtraction of a large 'blank value' from the sea waters, but this correction does not allow for any ammonia formed from organic nitrogen compounds in solution.

In spite of the errors accruing to this method, Raben's analyses of waters collected during the *Gauss* and *Planet* Expeditions showed that the deep water of the Atlantic contained very much more nitrate than the upper layers.

Table XIV. Nitrate nitrogen and ammonia nitrogen per cubic metre, mean value from a number of samples collected in the open ocean by the *Planet*.

Depth in metres	Temperature ° C.	Nitrate N_2 (mg. per cu. m.)	NH_3-N_2 (mg. per cu. m.)
0	22·0	101	49
400	9·6	313	47
800	6·2	485	45
1000	4·4	461	30
1500	—	271	38
2000	2·6	496	47
3000	2·5	510	107

The tropical West Pacific Ocean was found to be poorer in nitrate than the Atlantic Ocean in all the layers investigated.

From a series of analyses of water in the North Sea collected throughout the year Brandt and Raben found a seasonal variation in the nitrate owing to its utilisation by diatoms and algae in the summer months.

McClendon ((42) 1918) carried out estimations with water collected in the Caribbean Sea, using a similar method. He found less than 10 mg. nitrate + nitrite nitrogen per cubic metre in the water at or close to the surface, and attributed this low value as compared with those of Raben in the North Sea to the denitrifying

Table XV. Combined nitrogen content of surface waters—North Sea. Brandt and Raben, quoted by Krummel.

	Milligrams per cubic metre	
	N_2 as ammonia	N_2 as nitrite and nitrate
1904, Feb.	63	216
May	65	217
Aug.	61	79
Nov.	44	101
1905, Feb.	61	201

action of *Pseudomonas calcis*. This bacterium had been found in the water by Drew, who also found that it was capable of converting calcium nitrate into calcium carbonate in culture media. It is much more probable that the low value is due to the utilisation of nitrate by phytoplankton.

The writer ((9) 1926) has utilised a reduction product of strychnine, which Denigès had found to be a very delicate reagent for the detection of nitrites and nitrates, in order to obtain a quantitative measure of the nitrate in sea water, nitrite not being present in sufficient quantity materially to affect the result in the case of unpolluted waters from the open sea. This compound, dissolved in sulphuric acid, provides a reagent which gives a red colour with the small quantities of nitrate occurring in sea water, the depth of colour being directly proportional to the nitrate present up to a certain concentration of nitrate, above which the colour changes to orange and finally to yellow.

It was found that the upper layers of water from the English Channel during the summer months gave only a very faint tint on the addition of an equal quantity of this reagent, whereas with the same sample of water to which nitrate containing 10 mg. of nitrate-nitrogen had been added, a decided pink colour was obtained. Judging by the length of the columns which gave the same depth of colour, frequent instances were found where the upper layers contained no more than 2 or 3 mg. of nitrate-N_2 per cubic metre. Very low concentrations were also found in the upper layers of the Atlantic in the tropics, and at a number of positions in temperate latitudes during the summer.

The method provides good rough quantitative data, and since the differences found are very large, such data are sufficient to give a picture of the distribution of nitrate in the sea.

A regular seasonal variation occurs in the nitrate content of the water of the English Channel. Some 60 to 70 mg. of nitrate nitrogen per cubic metre is present throughout the column of water from top to bottom during the winter months, rapidly decreasing in the upper layers in the spring and being nearly all used up throughout all depths by August, after which it is re-formed in the bottom water faster than the plants use it (see Fig. 62, p. 179 and Table XVI).

Table XVI. Seasonal variation in nitrate content of water. English Channel. Lat. 50° 02′ N., long. 4° 22′ W.

	Milligrams of nitrate nitrogen per cubic metre of water	
	In upper water, 5 metres	In bottom water, 50–70 metres
1925, April 22	2	24
May 13	2	8
June 3	3	9
July 8	4	4
Aug. 5	Not detectable	Not detectable
Aug. 31	Not detectable	Not detectable
Oct. 1	6	(32)
Nov. 11	68	(70)
Dec. 11	38	43
1926, Feb. 3	83	64
Mar. 11	68	78
April 10	75	76
May 17	4	6
July 8	<5	10

The same method showed a considerable concentration of nitrate in the deep water of the Atlantic ((7a) 1926) and in the northern part of the North Sea.

The nitrate content of the water in Plymouth Sound, into which rivers and the sewage of a large town drain, is of particular interest because it shows that the plant growth in the rivers and estuary uses up most of the nitrate before it reaches the open sea during the summer months. (Table XIX.)

Table XVII. North Atlantic (west of Portugal). October 12, 1925. Lat. 37° 44′ N., long. 13° 21′ W.

Depth in metres	Temp. ° C.	Nitrate-N_2, mg. per cu. m.
0	21·10	15; 11
10	21·10	7
20	21·00	6
30	21·00	—
40	21·00	16
50	20·01	*ca.* 6
75	17·31	6
100	15·10	55
150	15·06	65
200	13·86	100
300	12·25	178; 158
500	10·94	200
1000	9·55	264; 262; 274
2000	4·81	*ca.* 265
3000*	3·10	*ca.* 265

* Oct. 16, 1925, lat. 29° 59′ N., long. 15° 03′ W.

Table XVIII. North Sea (south-west of Norway). May 16, 1925. Lat. 57° 57′ N., long. 6° 45′ E.

Depth in metres	Nitrate-N_2, mg. per cu. m.
0	6, 4
20	16
40	77
100	138, 128
300	160

Table XIX. Nitrate content of water in Plymouth Sound.

	Milligrams of nitrate-nitrogen per cubic metre of water	State of the tide
1925, May 13	(14)	1 hour after high water
Aug. 31	9	Low water
Oct. 1	112	5 hours after high water
Nov. 11	176	2 hours before high water
1926, Mar. 11	>190	1½ hours after high water
April 10	135	Low water
May 17	11	High water
July 8	24	5 hours after high water
Aug. 16	20	1 hour before high water
Sept. 22	17	2 hours after high water

In the vicinity of the submarine ridge which runs from Iceland to the Shetlands and occasions mixing of the deep water with the surface layers, a considerable concentration of both phosphate and nitrate was found in the upper layers during the middle of the summer.

Table XX. Nitrate and phosphate content of the water on the Wyville Thomson ridge in summer.
Collected in Channel between Faeroes and Shetland.

July 6, 1925. Lat. 61° 27′ N., long. 4° 23′ W.

Depth in metres	Nitrate-N_2, mg. per cu. m.	P_2O_5, mg. per cu. m.
10	67	30*
60	160	53
300	160	—
900	160	43
1000	160	58

July 6, 1925. Lat. 61° 02′ N., long. 3° 22′ W.

Depth in metres	Nitrate-N_2, mg. per cu. m.	P_2O_5, mg. per cu. m.
10	18	15*
40	78	33
100	92	43
500	106	54
800	115	59

North-west of Faeroe Islands.
July 4, 1925. Lat. 62° 53′ N., long. 9° 05′ W.

Depth in metres	Nitrate-N_2, mg. per cu. m.	P_2O_5, mg. per cu. m.
0	85	70*
10	70	30
30	115	30
80	Over 200	53
100	,,	60
300	,,	57
495	,,	59

* These values, particularly in the upper layers, may be high owing to phosphates being liberated from contained organisms after the samples were collected and before the analyses were made.

It is noteworthy that the proportion of nitrate-nitrogen to phosphate as P_2O_5 is similar both in the depths of the Atlantic and in the water of the English Channel in mid-winter. (An exception occurs in the water over the Wyville Thomson ridge.) Presumably not only the decay of plankton organisms gives rise to these products in the above proportion but also the requirements of the phytoplankton during life are in the same proportion, since both are almost completely used up by the phytoplankton in the upper layers of the sea during the summer.

A close comparison of the results so far obtained indicate that sometimes lack of phosphate and sometimes lack of nitrate is the immediate limiting factor for plant growth during the summer months in the English Channel (Fig. 62). However too much stress cannot be laid on small differences representing parts per thousand million, particularly as the estimation of the nitrate only leads to a rough approximate result, certainly sufficient to give a general picture of the round of events occurring in the sea, but not sufficiently accurate to draw very fine distinctions.

Ammonium Salts

Ammonium salts occur in sea water. Analyses of waters over an extensive area in the North and South Atlantic were made during the cruise of the *Planet*. From these Brandt ((6) 1916–1920) has drawn a diagrammatic section north and south through the Atlantic showing the amount of ammonia found in the water.

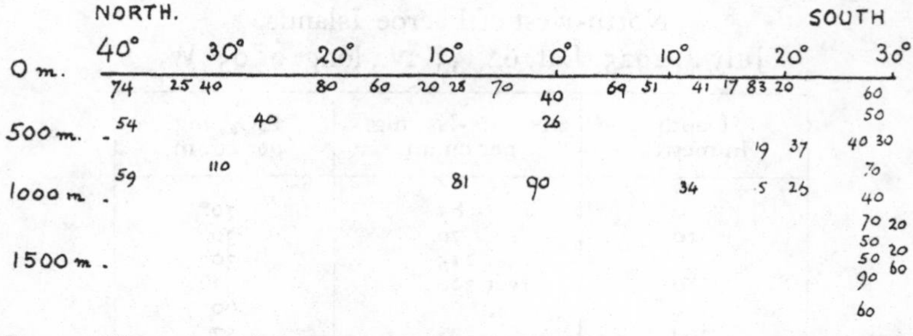

Fig. 4. Vertical section north and south through the Atlantic showing the concentration of ammonia (in mg. NH_3 per cubic metre) from observations made during the expeditions of the *Planet* and *Gauss*. (After Brandt, 1916–1920.)

Brandt and Raben have noticed no marked seasonal change in the ammonia of the North Sea.

Vernon ((8) 1899) found a considerable falling off in free ammonia in the Mediterranean on passing seaward from the shore.

Table XXI.

Distance from shore, at which water was collected. Kilometres	Free NH_3 mg. per cu. m.	Albuminoid NH_3 mg. per cu. m.
1	54	105
2	15	94
3	10	62
5	8	72
15	4	71

He made a number of experiments in order to gain some information concerning the causes of variation in this constituent, and noticed that there was a decrease in the free ammonia on keeping water from the Naples Aquarium, either in the dark or in diffuse light, with consequent growth of plant life. Open sea water, collected five kilometres off the shore and kept in complete darkness, showed variations in free ammonia, presumably the result of bacterial activity.

Similar experiments have been carried out by Stowell with water from the Bay of Biscay and from the aquarium of the Zoological Society of London ((34) 1926). In each case there was a falling off in free and in albuminoid ammonia during storage in the dark, while after exposure to sunlight with access to air and consequent growth of plant life there was in most cases an increase in both forms. These results indicate that storage of sea water in the dark, after filtration and aeration, is a very effective method of purification for aquarium purposes.

Nitrites

Nitrites are found in traces in some sea waters which have been stored, and in greater quantity in water which has been densely colonised with fish and invertebrates, such as the water of an aquarium. They are readily detected in traces of less than 1 milligram per cubic metre of nitrite-nitrogen by warming the water with a solution of α-naphthylamine and sulphanilic acid in glacial acetic acid ((9), (10) 1926).

An analysis of a number of samples of water from the Thames Estuary which were preserved from bacterial action by the addition of mercuric chloride when collected showed 1–2½ mg. nitrite-N_2 per cubic metre ((11) 1924). Orr has carried out estimations in estuarine water in the Clyde area, finding 1–10 mg. nitrite-N_2 per cubic metre, more being found in the deeper than in the surface water ((10) 1926).

In a number of samples of water from the open sea the writer has been unable to detect more than the minutest traces, at the most 1–2 mg. nitrite-nitrogen per cubic metre, even in water at 1000 and 3000 metres from the Atlantic. Raben found 'vanishingly small' traces of nitrite in numerous surface waters, and rather more in bottom water samples ((35) 1915).

Silicates

The water of the open sea contains a small quantity of silicate in solution. A seasonal variation has been observed in the English Channel and in the Baltic owing to its utilisation by diatoms; in the former area a spring maximum of 200–240 mg. SiO_2 per cubic metre has been ascertained to fall to 40 mg. or less. The deep water of the Atlantic contains silicate in greater abundance, values of 360 mg. SiO_2 per cubic metre having been found at 1000 metres and 1200 mg. at 3000 metres, the surface water containing 100–200 mg. per cubic metre. River and estuarine waters are relatively rich in silica.

Brandt has suggested that lack of silica may limit diatom growth, but the data do not bear this out in any case observed as yet.

Silicates in solution in water may be estimated by adding 2 c.c. of a 10 % ammonium molybdate solution to 100 c.c. of water and then four drops of 50 % sulphuric acid when a yellow colour develops, reaching a maximum in 10 to 20 minutes, after which it remains constant for some time. The depth of colour is a linear function of the silicate content, and may be matched against standards coloured with picric acid. A solution of 37–40 mg. of picric acid per litre has a colour corresponding with that given by 50 mg. of SiO_2 per litre as silicate after addition of molybdate and acid ((33), (37) 1920–1926).

ELEMENTS OCCURRING IN TRACES

In addition to the salts already considered, a large number of elements occur in small traces. Small quantities of these occur in certain marine animals where they may have an important function, such as that of copper in the blood of invertebrates, of iron in plants and animals, and of arsenates perhaps in plants.

Aluminium. Easily detected (1).

Arsenic. 10 to 80 mg. per cubic metre (Gautier, *Compt. Rend.* **137**, 232, 374, 1903).

30 mg. per cubic metre in four samples collected off the Essex coast (*Fish. Invest.* II, **6**, No. 4, 1924).

25 mg. per cubic metre in water in English Channel (*Fish. Invest.* II, **6**, No. 3, 1924).

Found in water off the Azores, the quantity increasing with depth (Gautier and Clausman, *Compt. Rend.* **139**, 101).

Found in oysters and other marine animals.

It has recently been suggested that arsenates may partly replace phosphates in the metabolism of marine plants, as they have been found capable of doing in the case of the fresh-water algae *Stichococcus flaccidus* and *Spirogyra crassa* (Atkins and Wilson, *Journ. Mar. Biol. Assn.* **14**, 609–614, 1927; Comère, J., "Action of Arsenates on the Growth of Algae," *Bull. Soc. Bot. France*, **56**, 147–151).

Barium. Less than 200 mg. per cubic metre in water of English Channel (*Fish. Invest.* II, **6**, No. 3, 1924).

Boron. Found in sea water (Veatch, *Proc. Cal. Acad. Sci.* **2**, 7).

Caesium. Detected (Sonstadt).

Cobalt. Occurs in lobsters, 2 mg. per kilo, and in mussels, 0·136 mg. per kilo (Bertrand and Machlebœuf, *Compt. Rend.* **180**, 1993, 1925).

Copper. Repeatedly detected (Dieulafait, *Ann. Chim. Phys.* 5 série, **18**, 359, 1879).

Occurs in the haemocyanin of the blood of many marine invertebrates.

Fluorine. 300 mg. per cubic metre in Atlantic water (Gautier and Clausman, *Compt. Rend.* **158**, 1631, 1914).

Gold. Repeatedly observed, probably variable in quantity up to *ca.* 5 mg. per cubic metre (Wagoner, *Trans. Amer. Inst. Mining Eng.* **31**, 807, 1901). At present a subject of investigation at the Kaiser Wilhelm Institut.

Iodine. 20 to 2800 mg. per cubic metre have been recorded (Winkler, *Zeit. Anorg. Chem.* **5**, 29, Pt 1, 205, 1916).

It is present in marine algae in readily decomposable organic combination, and easily diffuses out from dead algae (Okida and Eto, *Journ. Coll. Agric. Imp. Univ. Tokio*, **5**, 341, 1916).

Iron. Detected in sea water after oxidation by boiling with bromine water (11), or after evaporation with nitric acid. No reaction is given for ferric iron by sea water to which acid and iron free hydrogen peroxide has been added (9). It is probably present as organic compounds. These act catalytically in hastening the oxidation of various easily oxidised substances and of hydrogen peroxide. The catalytic effect is inhibited by the presence of organic matter. ((12) 1925. See also Wattenberg, "Die Deutsche Atlantische Exped. auf *Meteor*," 4 Bericht, *Zeit. der Gesell. für Erdkunde*, Nr. 5/6, 1927, p. 308.)

Lead. Found in the ash of marine organisms (1).

Lithium. Detected (Dieulafait, Bizio).

Manganese. Easily detected (Dieulafait, *Compt. Rend.* **96**, 718, 1883). Occurs in the ash of seaweeds.

Nickel. 2 mg. per kilo dry weight found in molluscs (Bertrand and Machlebœuf, *Compt. Rend.* **180**, 1380, 1925).

Radium. Sea water is radioactive to a degree comparable with a content of 17×10^{-6} mg. of radium per cubic metre (Joly, *Radioactivity and Geology*, 45–48, London, 1909).

$1 \cdot 2 \times 10^{-6}$ mg. per cubic metre (Lloyd, *Am. Journ. Sci.* 4th Series, **39**, 580, 1915).

Rubidium. *ca*. 0·04 % found in many analyses (1).

Silver. Repeatedly observed. 10 mg. per cubic metre have been found by Malagati (*Ann. Chim. Phys.* 3rd series, **28**, 129, 1850).

Strontium. The skeleton of the Radiolarian, *Podecanelius*, consists almost entirely of strontium (Bütschli, *Deut. Süd-Polar Exped.* **9**, 237, 1908).

Vanadium. Found in the blood of an ascidian (Henge, *Zeit. Physiol. Chem.* **86**, 340, 1913), and in a holothurian (Phillips, *Am. Journ. Sci.* 4th series, **46**, 473, 1918).

Zinc. Detected by Dieulafait. Less than 100 mg. per cubic metre was found in the water of English Channel at a depth of 70 metres off the Wolf Rock (11).

DISSOLVED ORGANIC MATTER

Sea water contains a small amount of organic matter in solution as well as in the minute organisms which abound. Sea water having a hydrogen-ion concentration of $10^{-8 \cdot 1}$ grm. per litre or less will not absorb carbon dioxide from the atmosphere, yet, on storage, water from the upper layers of the English Channel becomes more acid and develops a scum of bacteria around the meniscus. Deep water generally shows little change in acidity and no perceptible scum. A similar increase in the hydrogen-ion concentration would be brought about by the breakdown into carbon dioxide of from 1·5 to 7·5 grm. of carbohydrate per cubic metre of sea water. The change in hydrogen-ion concentration is occasioned by the oxidation of dissolved organic matter and the breakdown of the constituents of organisms both in their respiration process and in the process of bacterial putrefaction. The change in hydrogen-ion concentration of water which has been freed from all contained organisms has not yet been investigated.

On heating sea water with an acid or alkaline solution of potassium permanganate some of the permanganate is reduced, the oxygen having oxidised organic matter. The amount of oxygen absorbed in oxidising the organic matter in different waters varies, as is seen in the following table. The actual value obtained depends largely upon the temperature and time during which they are heated, the strength of permanganate solution employed and upon the nature of the organic matter present ((41) 1921).

Table XXII.

Water from	Albuminoid NH_3, mg. per cu. m.	Unfiltered samples		Filtered samples	
		Milligrams per litre of oxygen consumed by			
		Acid permanganate	Alkaline permanganate	Acid permanganate	Alkaline permanganate
Burnham-on-Crouch (estuary)	180	1·00	1·28	0·80	1·10
Ipswich (tidal river)	160	0·98	1·15	0·76	0·96
West Mersea (Essex coast)	150	0·92	1·05	0·73	0·92
Whitstable (Kent coast)	140	0·74	0·85	0·55	0·64

Gran and Ruud ((43) 1926) have obtained values for the 'oxygen absorbed' from permanganate by the water of Oslo Fiord throughout an entire year. They found minimum values in the winter and maximum in the summer months, but the values were not proportional to the quantity of plankton present at the time. Little difference was found between unfiltered water and water from which the organisms had been filtered. The estimated quantity of organic matter in solution outweighed the quantity found present in living organisms. Data for the water of Kiel Fiord and the Baltic have also been obtained by Raben, using this method.

As the coast is approached from the open sea, the surface water tends to contain more organic matter filterable through a porcelain candle ((12) 1925).

Nitrogenous organic matter occurs in sea water in minute amounts; thus Brandt gives analyses of waters from the North Sea containing between 100 and 200 mg. of albuminoid ammonia nitrogen per cubic metre ((6) 1916–1920).

In the American fresh-water Lake Mendota the total organic nitrogen in solution ranged around 400 mg. per cubic metre and was found to contain *inter alia* the amino acids tryptophane, tyrosine and histidine to the extent of about 13 mg. per cubic metre and cystine about 4 mg. (Domogalla, Juday and Peterson, *Journ. Biol. Chem.* **63**, 269–285, 1925. Also Peterson, Fred and Domogalla, *ibid.* **63**, 287.)

Vernon ((8) 1899) made a large number of determinations of the albuminoid ammonia in sea water, mostly of aquarium tank water in which animals or plants were kept for a stated length of time. In a series of determinations of open sea surface water collected at varying distances from the shore, he found a falling off in albuminoid ammonia in the first two miles from the shore.

A striking result of these experiments was that the albuminoid ammonia increased in an aquarium water in which either an alga was kept or in which vegetable growth, etc., developed. However, no reference is made to the water being filtered efficiently before the determinations were carried out, hence it is not impossible that the increase was due to the greater number of minute organisms in the water.

A very real difficulty exists in any attempt to estimate the organic matter actually in solution. In order to filter off the minute organisms completely it is necessary to use a porcelain filter which will absorb some of the dissolved organic matter; centrifuging may be found to overcome this difficulty. Having freed the water from minute organisms, the organic matter cannot be estimated directly owing to the relatively enormous quantity of salts present from which it is inseparable.

DISSOLVED OXYGEN AND NITROGEN

The surface layers of the sea are nearly in equilibrium with the air in regard to both these gases; the weight, or volume at N.T.P., of dissolved oxygen per litre of water varies with the temperature of the water, but the pressure is usually in the neighbourhood of 159 mm., as it is in water saturated with air at normal barometric pressure.

Close inshore and in estuaries with a rich flora of seaweeds, a regular diurnal variation in the oxygen content of the water has been found both by Legendre at Concarneau and Arcachon, and by the writer near Plymouth. Jacobsen found a similar diurnal variation in the upper layers of the Atlantic and Mediterranean. The maximum occurs in the evening after a period of sunshine and photosynthesis by the phytoplankton and fixed algae ((13) 1912). During the summer months the water of shallow estuaries is often supersaturated with oxygen.

In the deep water lochs of the Clyde Sea Area, Marshall and Orr have found a close relation throughout a year between the dissolved oxygen in the upper layers and the growth of diatoms ((44) 1927).

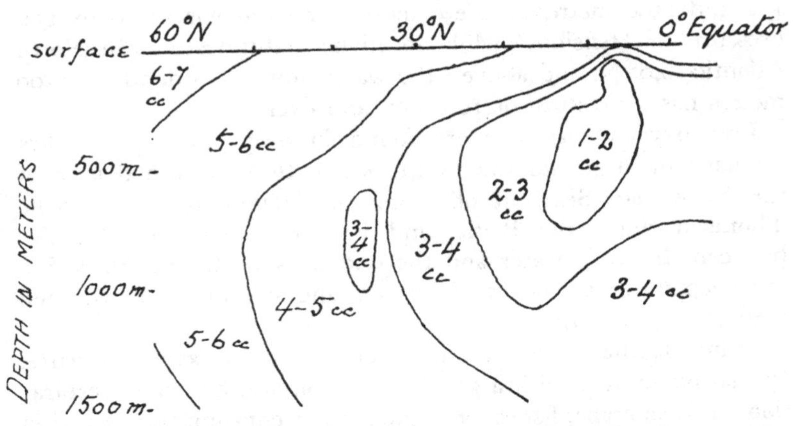

Fig. 5. Volume of dissolved oxygen in c.c. at N.T.P. per litre in water of the N. Atlantic between the equator and 60° N. (After Brennecke.)

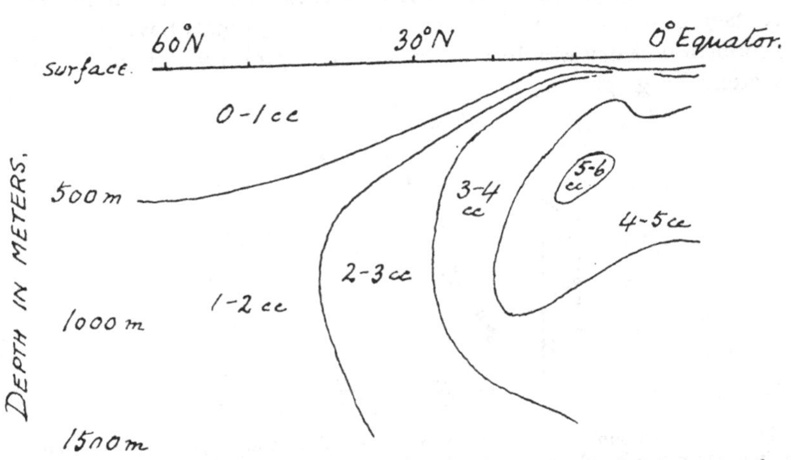

Fig. 6. Deficiency from saturation with oxygen in c.c. per litre in water of the N. Atlantic between the equator and 60° N. (After Brennecke.)

Brennecke ((14) 1909), from a large number of observations made on board the *Planet*, evolved a diagrammatic section north and south through the Atlantic showing the volume of dissolved oxygen at N.T.P. per litre of sea water and the deficiency from saturation in c.c. per litre (Figs. 5 and 6). The water between 100 and 1500 metres in the equatorial regions has a low oxygen pressure or 'tension,' while in higher latitudes in the North Atlantic—40° N. and above—the water down to a depth of 1500 metres has a pressure of 100 mm. and over.

Low oxygen pressures are found in deep water which lies stagnant in large hollows in the sea bottom, as the depths of the Norwegian Sea, cut off from the Atlantic by the Wyville Thomson ridge, the Baltic, and in fiords where a ridge lies between the deep water and the outside sea. In the Black Sea the deep water is devoid of oxygen, and sulphuretted hydrogen is present in quantity.

Animal life has been found in water of very low oxygen pressure. In marine animals with a good circulation and branchial respiration, such as crabs, fishes, etc., the oxygen consumption is within wide limits independent of its pressure in the water; this is also the case with hyaline pelagic organisms having a very small percentage of dry weight, but not with marine animals having a large proportion of dry weight and imperfect respiratory and circulatory systems (Fig. 7).

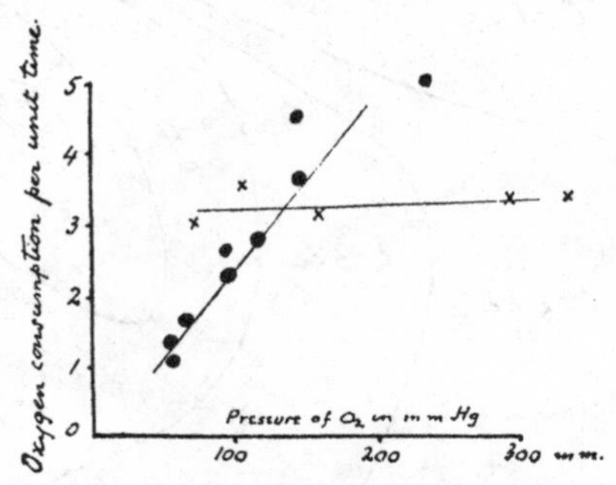

Fig. 7. Oxygen consumed in unit time by a crab and an anemone, from determinations made by Henze quoted by Krogh. Dots show values observed with *Anemonia sulcata*, and crosses show values for *Carcinus mænas*.

In warm seas, a larger volume of water will have to make contact with an animal's respiratory membranes, since the water contains less weight of oxygen per unit volume than in colder regions. In addition to this, at the higher temperature the animals will require more oxygen owing to their greater rate of metabolism.

Many marine animals, such as medusae, copepods and very young fish, do not survive when kept *in vitro* unless the water is nearly saturated with oxygen and also kept moving, so that there is a more or less continual flow past the organism. The reason for this is not very obvious; however, such movement will increase the efficiency of respiratory exchange and discourage the growth of bacteria on the surface of the animal A very ingenious apparatus to effect movement in a jar of water has been described by E. T. Browne ((15) 1897) and is shown diagrammatically in Fig. 8. It has been in continuous use and of

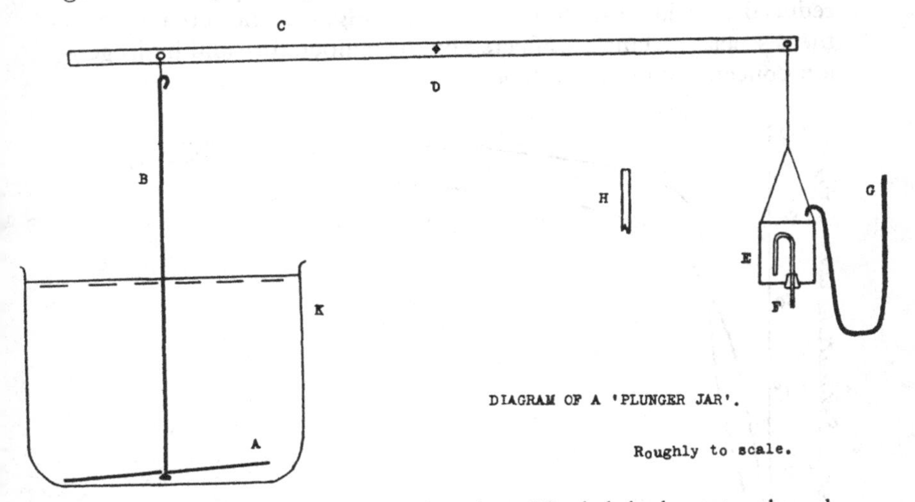

Fig. 8. The plunger *A* consists of a glass plate with a hole in the centre, through which passes a glass rod *B* suspended from a wooden rod *C*. The glass rod has a knob at the bottom end, upon which the glass plate rests, and is slightly bent so as to give a slope to the plate. The slope prevents small organisms being caught between the plate and the surface film when the plunger moves up. The wooden rod *C* rests near its centre on a pivot *D*, like the beam of a balance. At one end is suspended the plunger and at the other end a small tin can *E* fitted with a siphon *F*. A thin rubber tube *G* conveys a constant stream of water from a fresh water supply tap into the tin can which gradually descends as the weight of contained water increases until arrested by the stop *H*, when a small portion of the plunger should have just broken the surface of the water in the vessel *K*, an inverted bell jar. When full the tin can rapidly empties itself through the siphon and, being lighter than the plunger when empty, the plunger descends; the siphon must be of large bore so as to carry away the water faster than it comes in; the flow of water from the supply tap is regulated so that a complete cycle occupies about 80 seconds. The weights on either side of the pivot are regulated by a piece of lead which can be attached to the beam at the desired position.

considerable service in the Plymouth laboratory for a number of years. In fact it has provided the means of investigating the feeding habits of plankton organisms including the young stages of many economically important fishes.

The capability of the majority of marine animals to live for a time in water with a low oxygen pressure is understandable, since the blood contains either haemoglobin or haemocyanin and these blood pigments have the power of holding a considerable volume of oxygen in loose combination, and retain some 80 % of the oxygen they are capable of holding until the oxygen pressure is reduced to below *ca.* 30 mm. pressure (Fig. 9). The actual form of the dissociation curve is affected by the temperature and hydrogen-ion concentration of the blood (16).

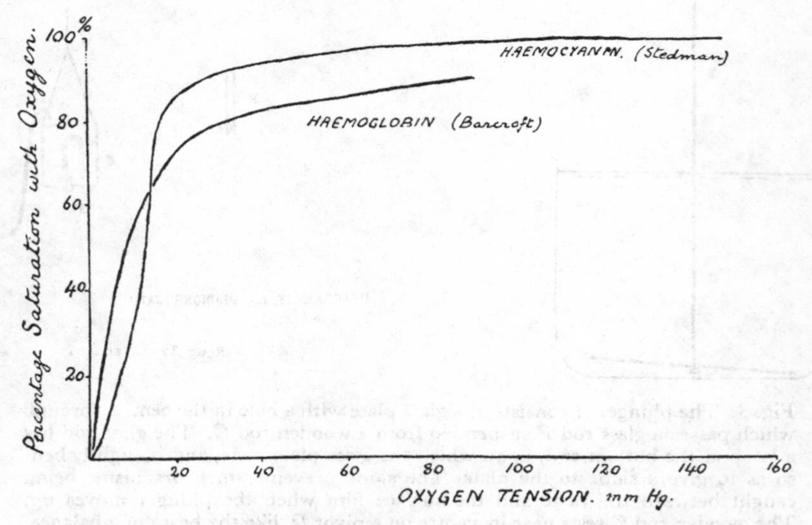

Fig. 9. Percentage saturation with oxygen of haemocyanin from the blood of a lobster at room temperature and of haemoglobin at 37° C. at various oxygen pressures.

A striking instance of the wealth of marine animal life in water of low oxygen content was found in the Gulf of Panama ((16a) 1925), where the water at a depth of 150–300 metres was only saturated with oxygen to an extent of 2 to 12 %. In the Caribbean Sea on the other side of the Panama isthmus the water between these depths was 50 to 63% saturated, yet in the low oxygen content water in the Gulf of Panama a much larger amount of animal life was found.

Table XXIII.

Approximate depth of horizontal haul with net. Metres	At a position in the Gulf of Panama, depth 3140 metres		At a position in the Caribbean Sea, depth 1500 metres	
	Quantity of plankton (c.c.) caught per haul	% saturation of the water with O_2	Quantity of plankton (c.c.) caught per haul	% saturation of the water with O_2
25	100	95	350	100
50	900	25	250	93
150	1000	10	250	63
300	1000	2	100	50

The volume of oxygen required to saturate sea water has been determined gasometrically with great accuracy by C. J. J. Fox ((17) 1907) who gives very complete tables of the solubility of both oxygen and nitrogen in sea water of various salinities.

The amount of oxygen dissolved in a small sample of sea water can be rapidly and accurately determined by Winkler's method ((18) 1921). This is conveniently carried out as follows:

The water is filled into a 50 to 100 c.c. glass-stoppered bottle, of known capacity when the stopper is in place, by means of a glass tube reaching to the bottom. In order to wash away any water which has been in contact with air during the operation, the water is allowed to overflow before the filling tube is withdrawn. Then 0·5 c.c. of 40% manganous chloride solution is run in with a pipette the point of which is kept near the bottom of the bottle in order that the solution may form a bottom layer. One c.c. of a solution containing 10 gm. NaOH and 5 gm. potassium iodide in 100 c.c. of water is then added in the same way and the pipette carefully withdrawn. The stopper is replaced in the bottle, care being taken that no air bubbles are included, and the bottle well shaken and allowed to stand until the precipitate of manganous hydroxide has settled. In order to minimise the risk of atmospheric oxygen entering past the stopper, the bottles may be kept covered with water while standing.

The white manganous hydroxide formed absorbs all the oxygen in the water in the bottle, turning to brown manganic hydroxide. Three or four c.c. of concentrated hydrochloric acid are poured into the neck of the bottle and the stopper replaced without air bubbles, care being taken not to disturb the precipitate more than possible. The bottle

Table XXIV. Number of c.c. of oxygen at N.T.P. dissolved in 1 litre of sea water saturated with air at the temperature shown and at 760 mm. pressure. The chlorine content of the sea water is given in grams Cl per 1000 gm. of sea water (Fox).

	Cl=0	Cl=1	Cl=2	Cl=3	Cl=4	Cl=5	Cl=6	Cl=7	Cl=8	Cl=9	Cl=10	Cl=11	Cl=12	Cl=13	Cl=14	Cl=15	Cl=16	Cl=17	Cl=18	Cl=19	Cl=20
0° C.	10·29	10·17	10·06	9·94	9·83	9·71	9·59	9·48	9·36	9·25	9·13	9·01	8·90	8·78	8·66	8·55	8·43	8·32	8·20	8·08	7·97
1° C.	10·02	9·90	9·79	9·68	9·57	9·45	9·34	9·23	9·12	9·01	8·89	8·78	8·67	8·56	8·44	8·33	8·22	8·11	8·00	7·88	7·77
2° C.	9·75	9·64	9·53	9·43	9·32	9·21	9·10	8·99	8·88	8·78	8·67	8·56	8·45	8·34	8·23	8·12	8·02	7·91	7·80	7·69	7·58
3° C.	9·50	9·39	9·29	9·19	9·08	8·98	8·87	8·77	8·66	8·56	8·45	8·35	8·24	8·14	8·03	7·93	7·82	7·72	7·61	7·51	7·40
4° C.	9·26	9·16	9·06	8·95	8·85	8·75	8·65	8·55	8·45	8·35	8·24	8·14	8·04	7·94	7·84	7·74	7·64	7·53	7·43	7·33	7·23
5° C.	9·03	8·93	8·83	8·73	8·64	8·54	8·44	8·34	8·24	8·14	8·05	7·95	7·85	7·75	7·65	7·56	7·46	7·36	7·26	7·16	7·07
6° C.	8·81	8·71	8·62	8·52	8·43	8·33	8·24	8·14	8·05	7·95	7·86	7·76	7·67	7·57	7·48	7·38	7·28	7·20	7·10	7·01	6·91
7° C.	8·60	8·50	8·41	8·32	8·23	8·14	8·04	7·95	7·85	7·77	7·68	7·59	7·50	7·40	7·31	7·22	7·13	7·04	6·95	6·85	6·76
8° C.	8·40	8·31	8·22	8·13	8·04	7·95	7·86	7·77	7·68	7·59	7·51	7·42	7·33	7·24	7·15	7·06	6·97	6·89	6·80	6·71	6·62
9° C.	8·21	8·12	8·03	7·95	7·86	7·77	7·69	7·60	7·52	7·43	7·34	7·26	7·17	7·09	7·00	6·91	6·83	6·74	6·66	6·57	6·48
10° C.	8·02	7·94	7·85	7·77	7·69	7·60	7·52	7·44	7·36	7·27	7·19	7·10	7·02	6·94	6·85	6·77	6·69	6·60	6·52	6·44	6·35
11° C.	7·84	7·76	7·68	7·60	7·52	7·44	7·36	7·28	7·20	7·12	7·04	6·96	6·88	6·80	6·71	6·63	6·55	6·47	6·39	6·31	6·23
12° C.	7·68	7·60	7·52	7·44	7·36	7·29	7·21	7·13	7·05	6·97	6·89	6·82	6·74	6·66	6·58	6·50	6·43	6·35	6·27	6·19	6·11
13° C.	7·52	7·44	7·36	7·29	7·21	7·14	7·06	6·98	6·91	6·83	6·76	6·68	6·61	6·53	6·46	6·38	6·31	6·23	6·15	6·08	6·00
14° C.	7·37	7·29	7·21	7·14	7·07	7·00	6·92	6·85	6·77	6·70	6·63	6·55	6·48	6·41	6·34	6·26	6·19	6·11	6·04	5·97	5·89
15° C.	7·22	7·15	7·07	7·00	6·93	6·86	6·79	6·72	6·64	6·57	6·50	6·43	6·36	6·29	6·22	6·14	6·07	6·00	5·93	5·86	5·79
16° C.	7·08	7·01	6·94	6·87	6·80	6·73	6·66	6·59	6·52	6·45	6·38	6·31	6·24	6·17	6·10	6·03	5·96	5·89	5·82	5·76	5·69
17° C.	6·94	6·88	6·81	6·74	6·67	6·60	6·54	6·47	6·40	6·33	6·26	6·20	6·13	6·06	5·99	5·93	5·86	5·79	5·72	5·66	5·59
18° C.	6·81	6·75	6·68	6·62	6·55	6·48	6·42	6·35	6·28	6·22	6·15	6·09	6·02	5·96	5·89	5·83	5·76	5·69	5·63	5·56	5·49
19° C.	6·69	6·63	6·56	6·50	6·44	6·37	6·30	6·24	6·17	6·11	6·05	5·98	5·92	5·86	5·79	5·73	5·66	5·60	5·53	5·47	5·40
20° C.	6·57	6·51	6·44	6·38	6·33	6·26	6·19	6·13	6·07	6·00	5·95	5·88	5·82	5·76	5·69	5·63	5·56	5·50	5·44	5·38	5·31
21° C.	6·46	6·40	6·33	6·27	6·22	6·15	6·09	6·03	5·96	5·90	5·85	5·78	5·72	5·66	5·59	5·53	5·47	5·41	5·35	5·29	5·22
22° C.	6·35	6·29	6·23	6·17	6·11	6·04	5·98	5·92	5·86	5·80	5·75	5·68	5·62	5·56	5·50	5·44	5·38	5·32	5·26	5·20	5·13
23° C.	6·24	6·18	6·12	6·06	6·01	5·94	5·88	5·82	5·77	5·71	5·65	5·59	5·53	5·47	5·41	5·35	5·29	5·23	5·17	5·11	5·04
24° C.	6·14	6·08	6·02	5·97	5·91	5·84	5·79	5·73	5·67	5·61	5·55	5·50	5·44	5·38	5·32	5·26	5·20	5·14	5·09	5·03	4·95
25° C.	6·04	5·99	5·92	5·87	5·81	5·75	5·69	5·64	5·58	5·52	5·46	5·41	5·35	5·29	5·23	5·17	5·12	5·06	5·00	4·95	4·86
26° C.	5·94	5·89	5·82	5·77	5·71	5·66	5·60	5·55	5·49	5·43	5·37	5·32	5·26	5·20	5·14	5·09	5·03	4·97	4·92	4·86	4·78
27° C.	5·84	5·79	5·73	5·67	5·62	5·57	5·51	5·46	5·40	5·34	5·28	5·23	5·17	5·11	5·06	5·00	4·94	4·89	4·83	4·78	4·70
28° C.	5·75	5·69	5·64	5·58	5·53	5·48	5·42	5·37	5·31	5·25	5·19	5·14	5·08	5·02	4·97	4·91	4·86	4·80	4·75	4·69	4·62
29° C.	5·66	5·60	5·55	5·49	5·44	5·39	5·33	5·28	5·22	5·16	5·10	5·05	4·99	4·93	4·88	4·83	4·77	4·71	4·66	4·60	4·54
30° C.	5·57	5·51	5·46	5·40	5·35	5·30	5·24	5·19	5·13	5·07	5·01	4·96	4·90	4·85	4·79	4·74	4·68	4·63	4·58	4·52	4·46

is then shaken, when the precipitate dissolves; the manganic chloride momentarily formed from the manganic hydroxide reacts with the potassium iodide and sets free iodine (8 gm. or 5598 c.c. of oxygen at N.T.P. set free 127 grams of iodine.)

It is then necessary to titrate the iodine set free with a solution of sodium thiosulphate (a 0·5 % or a 0·25 % solution of ordinary 'hypo'), using starch dissolved in boiling water as indicator. The precise strength of the thiosulphate solution is found either by titrating an iodine solution of known and similar strength ($N/50$ or $N/100$) using a similar quantity of starch indicator, or the volume of thiosulphate solution required to titrate the iodine set free by a sample of air-saturated water of known oxygen content is determined. It should be noted that if a vessel of water at a determined constant temperature is saturated by shaking with air, it should be allowed to stand for several hours at that temperature in order to attain equilibrium, otherwise it may be supersaturated. The solution of sodium hydroxide and potassium iodide used in this process must be free from nitrite, which can be determined by adding hydrochloric acid when it should not immediately turn starch solution blue. The thiosulphate solution loses strength on keeping, particularly during the first few days after being made up, hence it is necessary to determine its strength or make an estimation with a saturated water, within a reasonable time of carrying out the process with the water sample of unknown oxygen content.

The pressure of dissolved nitrogen in the surface layers of the sea is very nearly in equilibrium with the pressure in the atmosphere above. The amount dissolved in the deeper layers presumably indicates the temperature of that water when it occupied a position at the surface.

It is often desirable to gain some idea of the speed at which oxygen will be absorbed from the air. When certain conditions are fulfilled, Bohr's formula applies.

Bohr's invasion coefficient, γ, at the particular temperature (the number of c.c. of gas which enter 1 square cm. of surface in 1 minute when the pressure of this gas in the atmosphere is 760 mm. greater than its pressure in the water)

$$= \frac{\text{Vol. of gas entering surface in 1 min.} \times 760}{(Pa - Pw) \times \text{area of air water surface}},$$

Pa being the gas pressure in the air and *Pw* the gas pressure in the water.

When water is flowing past a bubble of oxygen Krogh found that γ for water at 37° C. = 0·07, provided that the bubble is small. With larger bubbles a smaller value was found, attributed by Krogh to incomplete renovation of the surface ((19) 1910). This indicates that Bohr's formula only holds when both water and air at the interface are being continuously and rapidly renewed. When standing still, as in a jar, the necessary conditions are not fulfilled and the speed of invasion is very much less; thus the writer found oxygen from the air entered a jar of sea water, 60 % saturated rising to 90 %, at a rate of *ca.* 0·00002 c.c. per square cm. per minute, the temperature being 15° C. The rate did not vary greatly from the beginning to the end of the particular experiment and showed no relation to the value of $Pa - Pw$.

Adeney and his co-workers ((20) 1926) measured the rate at which thin films of de-oxygenated water and salt solutions attained saturation when exposed to the air, and noted the effect of temperature upon the rate.

Table XXV. Rates of solution of oxygen by 0·05 cm. films of de-oxygenated water when uniformly exposed to the air.

Time in seconds to attain percentage saturation				Percentage saturation
0° C.	10° C.	20° C.	30° C.	
1·0	0·8	0·7	0·5	10
6·0	4·6	3·9	3·3	50
39·6	31·2	26·0	21·6	99

In the case of quiescent columns of de-aerated water fully exposed to the air, the dissolved atmospheric gases were not found to remain concentrated at the surface, but the columns became aerated gradually, owing to downward streaming brought about by lowering of the density due to evaporation from the surface. The rate of aeration is largely controlled by the rate of downward streaming. This is increased by

(*a*) The temperature of the air being lower than that of the water.

(*b*) The air being dry or of low humidity.

(*c*) Wind, which gives rise both to greater evaporation and to an enlarged area of exposed water surface when ripples are formed.

(*d*) Increasing the salinity of a solution up to about 15 ‰. Evaporation increases the density at the surface both by cooling and by rendering the concentration of salts greater. Above about 15 ‰ the preponderating effect of increasing salinity is presumably to reduce evaporation.

Since downward streaming does not keep the surface film renewed rapidly enough, natural conditions will not cause aeration of a body of water at a rate as great as that calculated from the rate of aeration of a 0·05 cm. film, as shown in Table XXV. This approximates to the maximum possible rate, except of course in cases where the air is bubbled through the liquid or otherwise mixed under pressure.

Experiments with quiescent columns of water showed that the rate of oxygenation was 56 to 245 times slower than that calculated from the rate of intake by such a film, the value 56 being for dry and 245 for nearly saturated air. Experiments with columns of sea and of fresh water in which the surfaces were kept ruffled by means of a stirrer below the surface, a simulation of wavelets in nature, showed that they absorbed oxygen one-half and one-quarter as fast respectively as the rate of intake of oxygen by a 0·05 cm. film.

The results of some interesting experiments upon the rate of entry of oxygen into sea water covered with a thin film of oil have recently been published by Roberts (36) in connection with the effect of oil pollution upon life in the sea. From these it is apparent that an oil film of a thickness likely to be produced will have little effect upon the oxygenation of the sea under conditions met with at sea.

CARBON DIOXIDE AND HYDROGEN ION CONCENTRATION

Sea water is slightly on the alkaline side of neutrality, and the carbon dioxide which it contains is partly in solution but mainly in combination, as bicarbonates and carbonates.

Water from the upper layers of the open sea is not far from being in equilibrium with the atmosphere with respect to this gas. Krogh determined the carbon dioxide pressure in a series of samples of water, collected between Canada and Scotland, by shaking the water with a small bubble of an inert gas, after which the bubble contains carbon dioxide at a partial pressure equal to its pressure or 'tension' in the water. Such a bubble will contain about 0·02 % of CO_2, whence the pressure in the water is *ca.* $\frac{0 \cdot 02 \times 760}{100}$ or 0·15 mm. The same method is applicable to the determination of the pressures of dissolved oxygen and nitrogen ((21) 1923). Krogh found that in general the carbon dioxide pressure of the surface waters of the sea was less than its partial pressure in the atmosphere above. Hence it is absorbed from the air, and the sea acts as a regulator of the carbon dioxide in the atmosphere. The rate at which the gas enters the water was calculated ((22) 1904) from Bohr's coefficient of invasion, γ, which is taken as 0·01 for carbon dioxide.

From this the quantity entering each square centimetre of the surface per year ($5 \cdot 25 \times 10^5$ minutes), when the difference in CO_2 pressure between the water and the air is 0·001 of an atmosphere, equals

$$5 \cdot 25 \times 10^5 \times 0 \cdot 001 \times 0 \cdot 01 \text{ or } 0 \cdot 525 \text{ c.c. per annum.}$$

The rate of invasion or evasion of carbon dioxide is very slow. A volume of sea water, either under or over saturated with CO_2, may remain with its surface exposed to the atmosphere for some considerable period before attaining equilibrium.

The amount of carbon dioxide which could be dissolved by the ocean from the atmosphere (containing 0·03 % CO_2) would be very small if the water were neutral—not more than 0·49 c.c. per litre at 0° C. Actually the water is alkaline and contains 45 c.c. or more CO_2 per litre, which is present in the form of CO_2 in true solution, of undissociated H_2CO_3, of bicarbonates, $'HCO_3$ ions, and almost undissociated carbonates.

The total quantity of CO_2 present in a sample of sea water may be determined by acidifying the water and boiling or pumping off the CO_2, and measuring this gasometrically or by passing it through a solution of barium hydroxide (40).

The bases in sea water are slightly greater than the equivalent of stable acid radicals; there is a portion of *excess base* or '*alkali reserve*' which is in combination with carbonic acid.

When a base in solution is shaken with air containing carbon dioxide at a definite pressure, an equilibrium is attained between the undissociated carbonate and bicarbonate, the carbonate ions, bicarbonate ions, the undissociated carbonic acid, the dissolved CO_2 in the solution, and the partial pressure of CO_2 in the air.

$$''CO_3 \rightleftharpoons 'HCO_3 \rightleftharpoons H_2CO_3 \rightleftharpoons CO_2 \text{ in solution} \rightleftharpoons CO_2 \text{ in the air.}$$

When carbon dioxide gas dissolves in a neutral aqueous solution, part exists as dissolved CO_2 and part as carbonic acid, H_2CO_3, which is very nearly undissociated. The experimental evidence of Thiel and Strohecker ((24) 1914) points to about ½ % of H_2CO_3 and 99½ % of dissolved CO_2, whereas Walter and Cormack ((25) 1900) conclude that the proportion is about equal in a neutral solution.

The amount of carbon dioxide dissolved and existing *in these two forms* is proportional to the CO_2 pressure in the gases in equilibrium with the liquid, and is affected by temperature and by the salts in the particular solution.

Table XXVI. α, absorption coefficient, or number of c.c. of CO_2 at N.T.P. which saturate 1 c.c. of liquid when the CO_2 pressure is 760 mm.

° C.	Pure water	NaCl solution 2·0 % Cl
0	1·713	1·483
6	1·377	1·207
12	1·117	0·977
18	0·927	0·827
24	0·780	0·710

Bohr, *Ann. der Physik u. Chemie*, 62 (1797), 614.

In equilibrium with this undissociated carbonic acid and consequently with the total dissolved CO_2, there also exists dissociated carbonic acid

$$H_2CO_3 \rightleftharpoons 'HCO_3 + H^{\bullet}$$

First dissociation.

$$'HCO_3 \rightleftharpoons ''CO_3 + H^{\bullet}$$

Second dissociation.

The equilibrium is in both cases affected by temperature and by other ions in solution. The addition of NaCl to the solution, for instance, increases the proportion of $'HCO_3$ to H_2CO_3 (27).

The proportion of carbonate ions to bicarbonate ions—that is, the proportion undergoing the second dissociation—is small, rising as the solution becomes more alkaline.

Warburg ((26) 1922) calculates the ratios in which $''CO_3$ will be in equilibrium with $'HCO_3$ in solutions of varying H ion concentration:

Table XXVII.

H' ion concentration in grm. per litre	$\frac{\text{conc. } ''CO_3}{\text{conc. } 'HCO_3}$
10^{-7}	0·001
10^{-8}	0·01
10^{-9}	0·1

Auerbach and Pick have measured the hydrogen ion concentration of mixtures of $NaHCO_3$ and Na_2CO_3; their results indicate that with a hydrogen ion concentration greater than $10^{-8·5}$ grm. per litre, $''CO_3$ ions do not exist in appreciable quantity, while with a hydrogen ion concentration less than $10^{-11·5}$ bicarbonate ions do not exist in appreciable quantity. A solution containing 0·1 mols per litre Na_2CO_3 and 0·1 mols per litre $NaHCO_3$ had a hydrogen ion concentration of $10^{-10·1}$ grm. per litre.

Hence it may be inferred that an inappreciable quantity of carbonate *ions* exist in natural sea water, which has a H ion concentration of around $10^{-8·2}$ grm. per litre.

On adding acid to sea water the excess base is wholly or partly neutralised, and carbon dioxide set free. It takes some time for the carbon dioxide to leave the liquid and the new state of equilibrium to be attained—as long as 24 hours bubbling with air. When sufficient acid is added to increase the H ion concentration to 10^{-4} grm. per litre, no appreciable quantity of CO_2 can exist in the solution as $''CO_3$, $'HCO_3$, or H_2CO_3. Hence the quantity of acid required to bring the hydrogen ion concentration of a sample of sea water to 10^{-4} grm. per litre, the CO_2 set free being got rid of by boiling, is a measure of the amount of *excess base*, which it contains.

The estimation of the *excess base* may be carried out in either of two ways. A convenient quantity of sea water, for instance 100 c.c., may be titrated with $N/100$ HCl using a convenient indicator, such as dibrom-o-cresol-sulphonephthalein, until the colour changes, when the CO_2 set free is boiled off and the titration continued with boiling after each addition of acid (30), (23*a*), (23). As an alternative a convenient quantity of the sea water may be titrated using methyl orange until the colour of the indicator is the same as its colour at the same concentration in pure distilled water saturated with carbon dioxide; in this case the effect of the CO_2 in the solution is taken as being the same in both cases ((39) 1926).

Open ocean water requires an amount varying from 23 to 26 c.c. of *N*/100 acid according to geographical position and season to bring 100 c.c. to a hydrogen ion concentration of 10^{-4} grm. per litre, the CO_2 set free being boiled off or its effect allowed for. That is to say the *excess base*, present in the form of bicarbonate and carbonate, is at a concentration of 0·0023 to 0·0026 normal. There is less CO_2, combined and free, present in sea water fresh from the open ocean than would be required for all the excess base to be present as bicarbonate. For instance a sea water of '26' excess base, that is 0·0026 normal, at a hydrogen ion concentration of $10^{-8.1}$ grm. per litre actually contains 48 c.c. of CO_2 (reduced N.T.P.) per litre. As only about 0·5 c.c. of this is in solution as H_2CO_3 and dissolved CO_2, the remaining 47·5 c.c. is in combination with the excess base. The conversion of all the excess base into bicarbonate would require 0·0026 × 22400 or 58·3 c.c. of CO_2 and into carbonate half this amount.

$$BOH + CO_2 = BHCO_3$$
$$2BOH + CO_2 = B_2CO_3 + H_2O.$$

A simple calculation shows that about two-thirds of the excess base is present as bicarbonate and the remaining third as carbonate[1]. From the evidence already presented it follows that the carbonate present in sea water is practically undissociated.

Wattenberg has shown that the bottom water of the South Atlantic contains a slightly greater ratio of excess base to total salts than the water lying between 300 and 3500 metres and attributes this to calcium carbonate being dissolved out from the bottom deposits. A lower ratio found in the surface layers is attributed to the utilisation of calcium carbonate by organisms. (*Die Deutsche Atlantische Exped. 'Meteor'*, 3 Bericht, p. 141. Berlin 1927.)

[1] The value 0·0026 normal is possibly high, because a little of the acid used in the titration will have reacted with the indicator and a little of the base will be in combination as silicates, phosphates, borates, etc. Hence one-third is a maximal value for the proportion of carbon dioxide present as carbonate. This conclusion is borne out since only one or two c.c. of *N*/100 acid are required to titrate 100 c.c. of sea water, using phenol-phthalein as indicator.

There is a further equilibrium between the amount of H˙ ions in sea water and the ″CO_3 and ′HCO_3 ions.

$$H^{\cdot} \rightleftharpoons {}''CO_3 \rightleftharpoons {}'HCO_3 \rightleftharpoons H_2CO_3 \rightleftharpoons CO_2.$$

The relation between the total CO_2 present in sea water of varying excess base and the hydrogen ion concentration is shown in Fig. 10, due to McClendon. The scope of the diagram has been

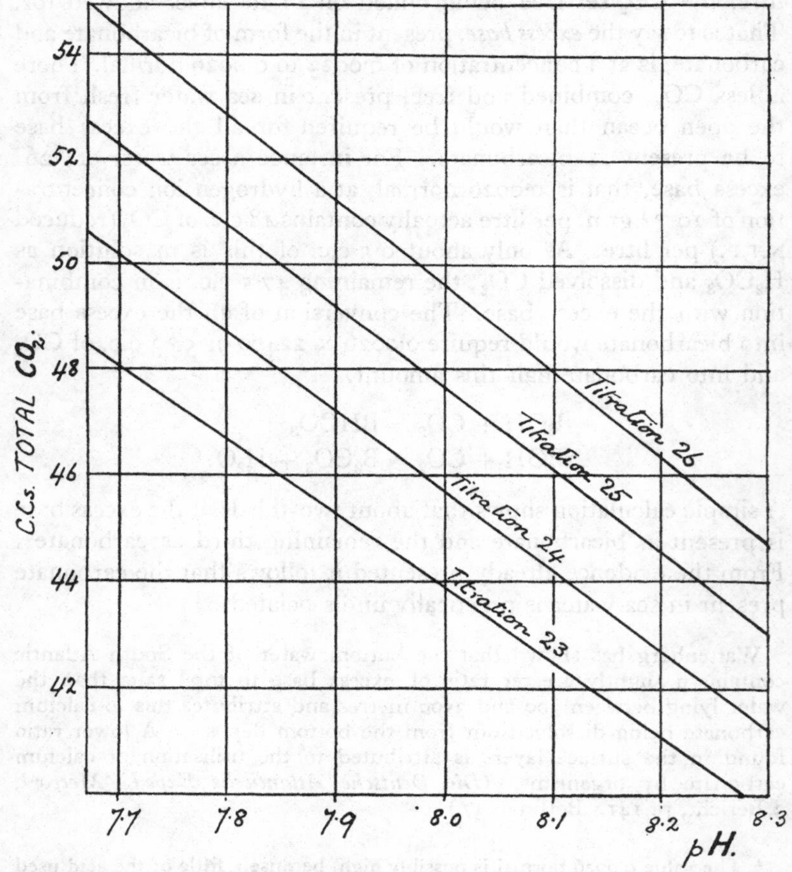

Fig. 10. Relation between the total CO_2 per litre and pH of sea waters of various 'excess base.' The titration values shown are the number of c.c. of $N/100$ HCl reacting with 100 c.c. of boiling sea water, using dibrom-o-cresol-sulphone-phthalein as indicator. The pH was not found to be affected perceptibly by temperature, provided there was no gain or loss of CO_2.

extended for waters of greatly varying salinity and for a wider range of hydrogen ion concentration by Bruce ((23 a) 1924).

It is customary to express hydrogen ion concentration of an aqueous solution in terms of pH, where

$$pH = \frac{1}{\log_{10} \text{hydrogen ion conc. in grm. equivalents per litre}}$$

Then $p\mathrm{H} = 7$ corresponds to 10^{-7} grm. of H ions per litre, or 0·000,000,1 normal. This is roughly their concentration in pure water, or any neutral aqueous solution where the number of H ions must equal the number of OH ions.

The hydrogen ion concentration of *pure* water cannot be estimated by adding an indicator, because indicators are themselves either acids or bases or salts and their addition to pure water will tend to make it more acid or alkaline, or, in the case where the indicator is a neutral salt its addition will tend to 'buffer' the water. In solutions, however, the salts, acids or bases present obliterate the effect of the minute quantity of indicator added.

In pure water or any solution the product of the grm. equivalents of H ions per litre multiplied by the grm. equivalents of OH ions per litre is known as the ionic product, K_W. It is the same whether the solution is acid or alkaline, but depends upon the temperature.

Table XXVIII.

Temperature (° C.)	Ionic product K_W
15	$0{\cdot}46 \times 10^{-14}$
20	$0{\cdot}86 \times 10^{-14}$
25	$1{\cdot}2 \times 10^{-14}$
30	$1{\cdot}8 \times 10^{-14}$

From this value, whatever the solution, we can calculate the concentration of the OH ions in grm. equivalents per litre, knowing the concentration of H ions. Thus if the H ion concentration of a solution at 25° C. is 10^{-4} grm. per litre the concentration of OH ions is $1{\cdot}2 \times 10^{-10}$ grm. equivalents per litre. Again the concentration of H ions in pure water or any neutral solution at 25° C. is $\sqrt{1{\cdot}2 \times 10^{-14}}$ grm. equivalents per litre.

In the same way $p\mathrm{H} = 8$ corresponds to 10^{-8} grm. of hydrogen ions per litre, which is on the alkaline side of neutrality, the OH ions being in excess of the H ions. As an instance, in such a solution at 25° C. the concentration of the OH ions will be $\frac{1{\cdot}2 \times 10^{-14}}{10^{-8}}$ or $1{\cdot}7 \times 10^{-6}$ grm. equivalents per litre.

$p\mathrm{H} = 6$ corresponds to 10^{-6} grm. H ions per litre, and the solution is on the acid side of neutral, the H ions being in excess of the OH ions.

The relation between the hydrogen ion concentration and the CO_2 pressure in a sample of sea water can either be calculated or found experimentally.

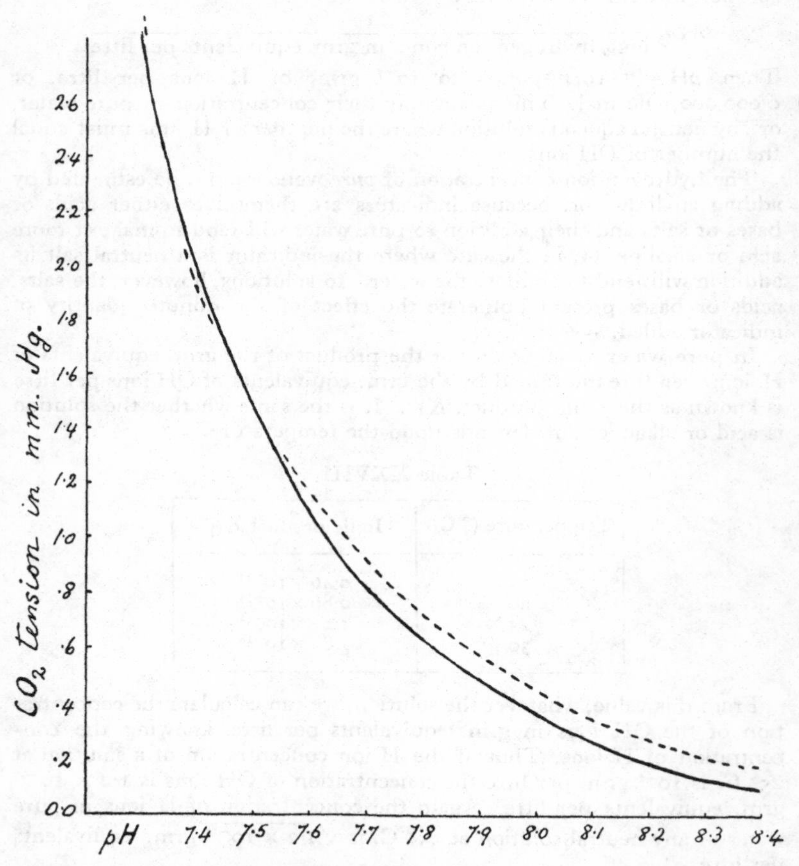

Fig. 11. Relation between *p*H and CO_2 pressure in sea water. Continuous curve from observations of McClendon at 20° C. on a number of sea water samples. Pecked curve from values calculated for a sea water of salinity 35 ‰ and excess base 0·0026 *N* from Saunders' equation.

Fig. 11 shows this relation as actually observed by McClendon for several open sea waters of somewhat varying salinity ((23) 1917). The relation between hydrogen ion concentration and the pressure of CO_2 in solutions containing bicarbonates and salts is of consequence in a great number of biological investigations. The *p*H may be measured in a very small quantity of the solution, whereas the CO_2 pressure, although readily measured by shaking a small quantity of inert gas with the liquid and finding the partial pressure of CO_2 in it, requires a greater quantity of liquid. If the solution contains salts or inert substances, such as cane sugar, the rate at which equilibrium is attained may be very slow. The relation in a solution of bicarbonate has been calculable from theory for several years. Saunders ((39) 1926) has extended this theory to solutions containing metallic cations in the form of neutral salts besides bicarbonates. The relation is shown by the equation

$$pH = 10{\cdot}70 - 0{\cdot}53\sqrt[3]{c} + \log\frac{[Bik]}{pCO_2},$$

where c is the concentration in terms of normality of all the cations including those derived from neutral salts, [Bik] is the concentration of excess base in terms of normality found by titration with methyl orange, and pCO_2 is the carbon dioxide pressure in mm. Hg. The equation is said to be valid provided c does not exceed 1·0.

The calculated CO_2 pressures for a sea water of 0·0026 excess base and salinity 35 ‰ is shown in Fig. 11, alongside the curve for values observed by McClendon. It is suggested that the somewhat lower pressures observed by McClendon at the higher *p*H values are due to CO_2 equilibrium not having been quite attained in the solutions (39).

From this figure it is apparent that sea water will absorb carbon dioxide from the atmosphere (atmospheric CO_2 pressure = *ca.* 0·23 mm.) when its H ion concentration is less than $10^{-8{\cdot}1}$ or $10^{-8{\cdot}2}$ grm. per litre, *i.e.* at greater *p*H values than *p*H 8·1 to 8·2, while at *p*H 8·1 to 8·2 it is approximately in equilibrium with the atmosphere. At higher H ion concentrations (lower *p*H values), it very slowly gives off CO_2 to the atmosphere.

Variations in temperature of a sample of sea water cause a small change in its hydrogen ion concentration. When no carbon dioxide is allowed to enter or leave the sample a rise of 1° C. decreases the *p*H by 0·01. If the water is warmed and carbon dioxide allowed to leave the sample as it becomes supersaturated with respect to this gas in the atmosphere, then the loss of CO_2 overcomes the effect of temperature alone, and the *p*H rises.

Sea water from the open ocean has a hydrogen ion concentration of between 10^{-8} and $10^{-8.3}$ grm. of hydrogen ions per litre, or *p*H 8·1 to 8·3.

During the summer months when the phytoplankton is assimilating CO_2, its amount in the water is lowered and the con-

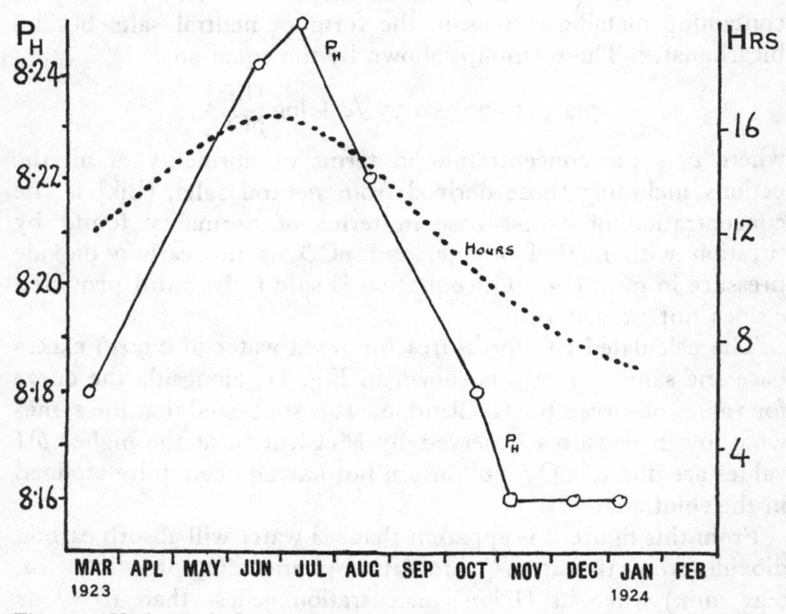

Fig. 12. Variation in hydrogen ion concentration of the surface water of the English Channel in terms of *p*H with the season and hours of daylight.

centration of hydrogen ions in equilibrium is likewise lowered (*i.e.* the *p*H value is raised). Under these circumstances a *p*H of *ca.* 8·3 may be attained in the upper layers. During winter months when respiration is in excess of assimilation, CO_2 in solution is greater and the *p*H falls to 8·1 in the upper layers, or less in deep and in estuarine waters. The effect of sunlight upon photosynthesis by phytoplankton and, in turn, upon the hydrogen ion concentration of the upper layers of the sea is very clearly shown in Fig. 12. In a rock pool with algae in strong sunshine a *p*H as high as *ca.* 8·6 may be attained, while with *Ulva* an even higher value, 9·7, has been noted.

Marshall and Orr have found a close relation between changes in the *p*H of the water in the Clyde Sea Area and increases in the number of diatoms present, an increase in *p*H taking place concurrently with a burst of diatom growth ((44) 1927).

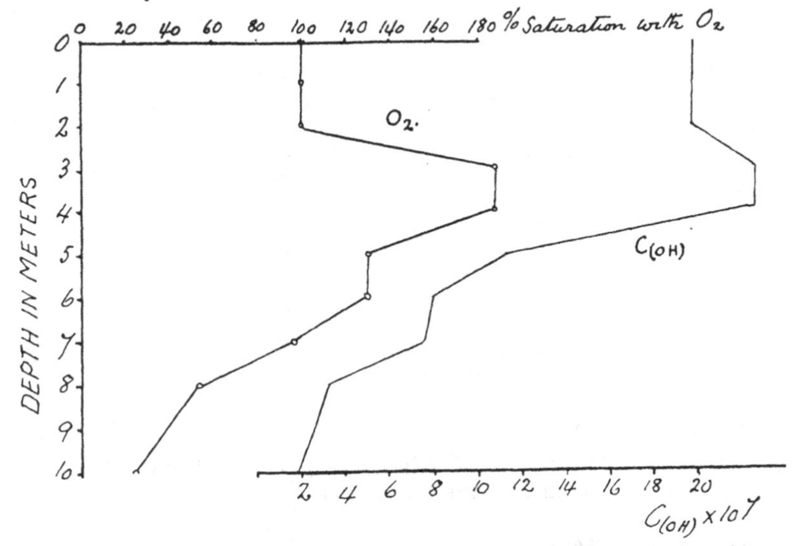

Fig. 13. Percentage saturation with oxygen and concentration of hydroxyl ions, C_{OH}, with depth in the 'Inderoepollen' (after Gaarder).

Gaarder ((28) 1916) has estimated the oxygen content and the hydrogen ion concentration of layers of water in a poll or basin, the Inderoepollen, which is connected with the open waters of a Norwegian fiord at its surface, while the water is more or less stagnant below. In Fig. 13 the percentage saturation with oxygen is shown together with the concentration of hydroxyl ions—which rises as the *p*H rises. The supersaturation of O_2 due to phytoplankton is clearly shown in the layer just below the surface. In the same layer the OH concentration is at a maximum, the dissolved carbon dioxide having been used up in photosynthesis. Below 7 metres the condition is due to excess of respiration over photosynthesis.

The following table shows values actually observed by Atkins in the places named. The values are not necessarily the actual maximum and minimum that occur.

Table XXIX.

	*p*H	Salinity ‰
In English Channel, surface, 20 miles S.W. of Plymouth breakwater	8·27–8·14	35·40–35·13
Off Plymouth breakwater	8·27–8·07	35·17–32·25
In Plymouth Sound	8·29–8·01	35·00–30·69
In shallow water	8·42–8·01	—
In a rock pool	8·57–8·01	—

Table XXX. Variation of hydrogen ion concentration with depth (Palitzsch).*

Depth in metres	N. Sea E. of Faroe Islands	N. Atlantic W. of Portugal	Mediterranean
0	*p*H 8·13	*p*H 8·22	*p*H 8·23
100	8·09	8·13	8·21
400	8·03	8·04	8·19
1000	7·98†	8·01	8·14
2000	—	7·95	8·09
3200	—	—	8·07

* All during summer months. † At 700 metres.

Table XXXI. Variation in hydrogen ion concentration with depth in the North Atlantic, at a position about 200 miles west of Portugal (Atkins 7*a*, 1926).

Depth in metres	Temp. °C.	*p*H
0	21·10	8·35
10	21·10	—
20	21·00	—
30	21·00	—
40	21·00	—
50	20·01	8·35
75	17·31	8·31
100	15·10	8·18
150	15·06	8·16
200	13·86	8·11
300	12·25	8·12
500	10·94	8·00
1000	9·55	8·03
2000	4·81	7·94
3000	3·10	7·87

When an alkali is added to sea water its hydrogen ion concentration is reduced and its hydroxyl ion concentration raised until a point is reached (*p*H 10) when it becomes supersaturated with respect to magnesium hydroxide, and magnesium and calcium carbonate. These separate out as a white precipitate; on the gradual addition of alkali the precipitate at first consists of magnesium hydroxide for the most part and later of the carbonates. When the precipitation is complete (at *p*H 11–12) a further addition of alkali causes a more rapid fall in hydrogen ion concentration ((38) 1926). If the hydrogen ion concentration is reduced gradually, the water may remain for a considerable period supersaturated with

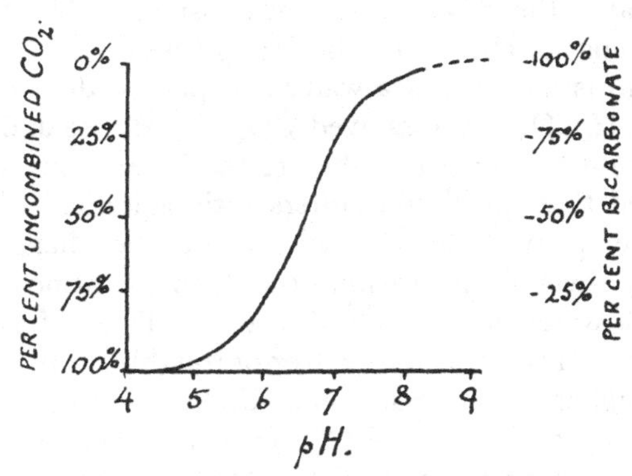

Fig. 14. Curve showing the equilibrium between hydrogen ions and uncombined CO_2 in a solution containing 1 % by weight of total CO_2, to which increasing quantities of NaOH have been added up to 0·909 % which is sufficient to combine with all the CO_2 to form bicarbonate. The dotted part of the curve shows the region in which a trace of Na_2CO_3 exists.

respect to bicarbonate ions, because the change $'HCO_3$ to $''CO_3$ proceeds at a very slow rate (39). It is tentatively suggested that this is the reason why the hydroxides are present in large amount in the precipitate formed by suddenly reducing the CO_2 pressure by the addition of alkali, whereas geological evidence indicates that calcium and magnesium carbonates have been deposited when the CO_2 pressure fell gradually in past ages. The localised areas of high OH ion concentration which occur when alkali is added also tend to cause the precipitation of hydroxide rather than carbonate, and the difference referred to may be attributed to this cause in some degree.

On adding acid to sea water, a *comparatively* large amount is required to bring about each increment of hydrogen ion concentration while bicarbonate exists in solution in appreciable quantity (down to pH 5·5). The sea water is 'buffered' by the bicarbonate it contains when of a pH between 5·6 and 8. Fig. 14 shows the concentration of H ions in a solution containing 1 % total CO_2 to which have been added increasing quantities of sodium hydroxide. It shows that a relatively large addition of alkali (or acid) is necessary to alter materially the pH of a solution buffered with bicarbonate between pH 5·5 and 8.

Both McClendon and Irvine (38) have shown that in addition to bicarbonates, the minute traces of phosphates, borates, silicates and organic matter also exert a buffering action.

When acid is added to sea water and part of the excess base is neutralised, H_2CO_3 and dissolved CO_2 are set free and these may remain present for some time before the latter is given off to the air, provided that the liquid suffers little agitation. The relation between the quantity of acid added and the change in pH is modified by the rate of evasion of CO_2 to the atmosphere; with a typical sea water of pH around 8, the addition of 1 c.c. $N/100$ acid to 100 c.c. of water causes a drop of roughly 0·1 pH if subjected to minimal agitation (30). Such an addition reduces the excess base by 1, but it does not set free CO_2 from bicarbonates in quantity equivalent to the quantity of acid added, because part of the acid combines with the carbonates. A water with excess base '26' and pH 8·25 contains 45·5 c.c. total CO_2, as seen from Fig. 10; on adding 1 c.c. $N/100$ acid to 100 c.c. of this water the excess base is reduced to '25' and the pH is reduced to 8·14 if no CO_2 is lost to the atmosphere. If on the other hand CO_2 is added to the 100 c.c. of water at pH 8·25, 0·180 c.c. will be required to lower its pH to 8·14.

DETERMINATION OF HYDROGEN ION CONCENTRATION

Very accurate determinations have been carried out by means of the indicator method, the values so obtained being checked electrometrically and showing close agreement.

Within the range of its colour change, the colour of any indicator depends upon the hydrogen ion concentration of the solution, and, in addition, but to a very much less extent, upon the amount and nature of salts in the solution. Proteins in solution also exert an effect but are not present in sufficient quantity in sea water to do so. Hence in sea water, containing sodium chloride, the depth of colour produced will not be the same as in solutions of other salts, such as phosphates or borates, with the same hydrogen ion concentration.

Table XXXII. McClendon's Buffer Solutions
(*Journn. Biol. Chem.* **30**, 280, 1917).

18·6 grm. boric acid & 22·5 grm. NaCl (dried in desiccator) made up to 1 litre with boiled distilled water c.c.	28·67 grm. borax crystals (dried on filter paper), 19 grm. NaCl (dried in desiccator) made up to 1 litre with boiled distilled water c.c.	pH of sea water giving equal colour on addition of aliquot amount of sulphonephthalein indicator		
		Salinity		
		26‰	33‰	39‰
23·9	6·1	7·5	7·45	7·4
22·8	7·2	7·6	7·55	7·5
21·6	8·4	7·7	7·65	7·6
20·4	9·6	7·8	7·75	7·7
19·2	10·8	7·9	7·85	7·8
18·0	12·0	8·0	7·95	7·9
16·8	13·2	8·1	8·05	8·0
15·6	14·4	8·2	8·15	8·1
14·1	15·9	8·3	8·25	8·2
12·6	17·4	8·4	8·35	8·3
11·1	18·9	8·5	8·45	8·4
9·6	20·4	8·6	8·55	8·5
7·8	22·2	8·7	8·65	8·6
6·0	24·0	8·8	8·75	8·7
4·2	25·8	8·9	8·85	8·8
2·4	27·6	9·0	8·95	8·9

In order to obviate this difficulty McClendon (23) prepared a series of standard solutions containing sodium chloride, the various members of which gave with one of the sulphonephthalein series of indicators a tint exactly equal to that given by one of a number of samples of sea water. The hydrogen ion concentration of the sea water samples was determined electrometrically.

From these determinations Table XXXII was developed. The standard or *buffer* solutions are so composed that a small addition of either acid or alkali has a minimum effect in altering their hydrogen ion concentration.

By using 0·50 c.c. of 0·02 % cresol red with 10 c.c. of sea water, and an equal quantity with 10 c.c. of each buffer solution, it was found possible to determine the H ion concentration to within *ca.* ± 0·01 *p*H (30). Standards intermediate in value between those shown in the table were also used. In each case the colour comparisons were made in hard glass test tubes of equal diameter—16 mm.—which were kept stoppered with rubber caps.

The buffer solutions are liable to alter on keeping owing to action on an unsuitable glass or to growth of moulds; this latter can be largely obviated by adding one drop of toluene to each tube of solution. These buffer solutions with the indicator and a drop of toluol added have been known to keep without change for several months in suitable hard glass tubes, stoppered and kept in the dark when not in use. A change taking place can often be seen by one or more tubes in a series falling out from the regular serial gradation of colour.

In order to assist in the colour comparison, there were used as a check, 0·50 c.c. of 0·02 % xylenol blue per 10 c.c. of sea water with waters having *p*H greater than *ca.* 8·2 and for sea waters below *ca.* 7·9 an equal quantity of 0·02 % phenol red.

In various biological experiments measurements are required of the hydrogen ion concentration, of acidified sea water of lower *p*H values than those of McClendon's series of buffer solutions of sea waters which have been diluted. When buffer solutions such as those of Sorensen or Clarke and Lubs are used a correction must be applied for the effect on the particular indicator of the neutral salts, chiefly NaCl, in the sea water. From Table XXXIII an approximate value of the '*salt error*' to be applied when using various indicators may be obtained. Open ocean water is *ca.* 0·55 *N* in respect to NaCl. A definite relation has been shown to exist between the 'salt error' of the sulphonephthalein indicators and the difference in normal concentration of the cations in the buffer mixture and in the solution under examination (39). A graphical representation of this relation makes it simple to calculate the 'salt error' to apply when using any particular buffer solution and salt solution of unknown *p*H.

The effect of temperature on the usual indicators is slight, that is to say the *p*H registered for a hot solution is very close to its *p*H as measured electrometrically at that particular temperature, subject of course to 'salt' and protein error. Legendre (31) and Saunders (39) give a list of corrections for various indicators due to temperature. The effect of temperature on the actual hydrogen ion concentration of sea water or salt solution is quite a different matter and, as already stated, amounts to a decrease in *p*H of 0·01 per 1° C. rise in temperature.

Table XXXIII. Clarke and Lubs' range of Indicators.

Concentration of indicator solution	Indicator	*p*H range of colour change	Salt solution	Correction for salt error
0·04 %	Thymol blue	1·2–2·8	0·1 *N* KCl	−0·08
			0·25 ,,	−0·08
			0·5 ,,	+0·02
			1·0 ,,	+0·23
			NaCl has about the same influence	
0·04 %	Bromphenol blue	3·0–4·6	0·1 *N* KCl	−0·05
			0·25 ,,	−0·15
			0·50 ,,	−0·35
			1·0 ,,	−0·35
0·02 %	Methyl red	4·4–6·0	0·5 *N* NaCl	+0·10
0·04 %	Bromcresol purple	5·2–6·8	0·5 ,,	−0·25
0·04 %	Bromthymol blue	6·0–7·6		
0·02 %	Phenol red	6·8–8·4	0·5 ,,	−0·15
0·02 %	Cresol red	7·2–8·8	Ocean water	−0·18
0·04 %	Thymol blue	8·0–9·6	0·5 *N* NaCl	−0·17

The ranges of three other common indicators are:

Litmus... ...	*p*H 4·5–8·3
Phenolphthalein	8·3–10
Methyl orange	3·1–4·4

BIOLOGICAL EFFECTS OF HYDROGEN ION CONCENTRATION

In the fish hatchery at Port Erin plaice were found liable to ulcerated areas on their skin when in water of high hydrogen ion concentration occasioned by numerous green algae and flagellates.

From observations in Puget Sound, an area having water of very varied hydrogen ion concentration, Shelford and Powers concluded that active migratory fishes were sensitive to small changes. The evidence for this conclusion is flimsy. It is noteworthy that a marked increase in rate of respiration of fish is seen on adding a trace of acid to sea water.

In the same area Gail found that the hydrogen ion concentration was a definite factor governing the distribution of *Fucus evanescens*.

In general it would appear that the relatively small differences which occur in the hydrogen ion concentration in the open sea do not play any well-marked part in regulating the fauna.

Where, however, animals are kept in sea water under conditions which do not conform to those in the open sea, and where the *p*H passes outside the range met with in nature, then profound effects become apparent. Biological problems centering upon the hydrogen ion concentration of sea water are discussed in a monograph by Legendre ((31) 1925) which contains many valuable references.

In addition to, and quite distinct from, the part played by varying concentrations of hydrogen ions upon living animals and plants, the measurement allows an accurate measure to be obtained of the quantity of carbon dioxide produced by the animals and plants in their respiration and used by the plants in photosynthesis.

SEA WATER AS A CHEMICAL MEDIUM FOR PLANT AND ANIMAL LIFE

From the foregoing account two noteworthy characteristics of sea water, as a medium for plants or animals, stand out clearly. It contains a readily available store of carbon dioxide, just as blood contains a store of oxygen. The store of carbon dioxide can be drawn upon or added to, and such alteration, if made within limits, occasions only a relatively slight shift in the concentration of hydrogen ions.

Marine animals are not only in contact with the sea water in which they live, but are readily permeable to the water molecules. The blood and body fluid of elasmobranch fishes and invertebrate animals are in osmotic equilibrium with sea water in nature, and if the osmotic pressure of the water is changed by dilution or concentration, their blood and body fluids rapidly attain almost the same osmotic pressure; in this respect marine animals appear different from lower fresh-water animals.

It is clear that marine animals are permeable to salts in solution in the sea water, but how freely permeable to each particular salt is as yet unknown. The salts in their blood and body fluids are in some cases very nearly in the same proportion as that in which they occur in the sea.

For plant life sea water is a nutritive medium containing all that is necessary for the requirements of algae, with the exception of nitrate and phosphate which are *on occasions* insufficient for a vigorous growth. It was thought at one time that a lack of silica at times also limited plant growth, but the more recent determinations by Atkins indicate that sea water is not depleted of this constituent to such an extent that it becomes a limiting factor.

BIBLIOGRAPHY

(1) CLARKE, F. W. *Data of Geochemistry*. Washington, 1924.

(2) OXNER and KNUDSEN. "Chloruration par la méthode Knudsen." *Bull. Comm. Internat. pour l'Expl. sci. de la Mer Méditerranéenne*, No. 3. April 1920.

(3) KNUDSEN, M. *Hydrographical Tables*. Copenhagen and London, 1901.

(4) EKMAN, V. W. *Publ. de Circonstance*, No. 43. Copenhagen, 1908. *Ibid.* No. 49. 1910.

(5) MATTHEWS, D. "On the Amount of Phosphoric Acid in the Sea Water off Plymouth Sound." *Journ. Mar. Biol. Assoc.* **9**, 121–30. 1916. *Ibid.* **9**, 251–75. 1917.

(6) BRANDT, K. "Über den Stoffwechsel im Meere." *Wiss. Meeresuntersuchungen*, **18**. Kiel, 1916–1920.

(7) ATKINS, W. R. G. "The Phosphate Content of Fresh and Salt Waters in its Relationship to the Growth of the Algal Plankton." *Journ. Mar. Biol. Assoc.* **13**, 119–50. 1923.

—— "Seasonal Changes in the Phosphate Content of Sea Water in Relation to the Growth of Algal Plankton during 1923 and 1924." *Journ. Mar. Biol. Assoc.* **13**, 700–20. 1925.

—— "The Phosphate Content of Sea Water in Relation to the Growth of the Algal Plankton." *Journ. Mar. Biol. Assoc.* **14**, 447. 1926.

—— "A Quantitative Consideration of some Factors concerned in Plant Growth in Water." *Journ. du Cons. Internat.* **1**, 197–226. 1926.

(7 *a*) ATKINS, W. R. G. and HARVEY, H. W. "The Variation with Depth of Certain Salts utilised in Plant Growth in the Sea." *Nature*, **116**, 784. 1926.

(8) VERNON, H. M. "The Relations between Animal and Vegetable Life." *Mitt. aus der Zoolog. Stat. zu Neapel*, **13**, 341–425. 1899.

(9) HARVEY, H. W. "Nitrate in the Sea." *Journ. Mar. Biol. Assoc.* **14**, 71–88. 1926.

(10) ORR, A. P. "The Nitrite Content of Sea Water." *Journ. Mar. Biol. Assoc.* **14**, 55–62. 1926.

(11) *Fisheries Investigations*, Series 2, **6**, No. 3. 1924.

(12) HARVEY, H. W. "Oxidation in Sea Water." *Journ. Mar. Biol. Assoc.* **13**, 953–69. 1925.

(13) JACOBSEN, J. *Report Danish Oceanograph. Expeditions*, 1908–1910. **1**, 209. Copenhagen, 1912.

(14) BRENNECKE, W. *Forschungsreise S.M.S. 'Planet,'* 1906–1907. **3**. Berlin, 1909.

(15) BROWNE, E. T. "On Keeping Medusae alive in an Aquarium." *Journ. Mar. Biol. Assoc.* **5**, 176–80. 1897.

(16) STEDMAN and STEDMAN. *Biochem. Journ.* **19**, 547. 1925.
BARCROFT, J. *The Respiratory Function of the Blood.* Cambridge, 1913.
PANTIN and HOGBEN. *Journ. Mar. Biol. Assoc.* **13**, 970. 1925.

(16 *a*) SCHMIDT, J. *Science*, **61**, 592. 1925.

(17) FOX, C. J. J. "On the Coefficients of Absorption of the Atmospheric Gases in Sea Water and Distilled Water." *Publ. de Circonstance*, No. 41. Copenhagen, 1907.

(18) JACOBSEN and KNUDSEN. "Dosage de O_2 dans l'Eau de Mer par la Méthode de Winkler." *Bull. de l'Institut Océanogr. Monaco*, No. 390. 1921.

(19) KROGH, A. *Scand. Arch. f. Physiol.* **23**, 224. 1910.

(20) ADENEY, W. E. "On the Rate and Mechanism of Aeration of Water under Natural Conditions." *Sci. Proc. Roy. Dublin Soc.* **18**, 211–18. 1926.

(21) HELLAND-HANSEN, B. "Ocean Waters." *Internat. Rev. f. Hydrobiol.* **11**, 449 et seq. 1923.

(22) KROGH, A. *Meddelelser om Gronland*, **26** and **31**. Copenhagen, 1904.

(23) MCCLENDON, J. F. *Journ. Biol. Chem.* **30**, 274. 1917.

(23 *a*) BRUCE, J. R. "A *p*H Method for Determining the Carbon Dioxide Exchanges of Marine Brackish Water and Freshwater Organisms." *Brit. Journ. Exp. Biol.* **2**, 57–64. 1924.

(24) THIEL and STROHECKER. *Ber. Deutsch. Chem. Gesell.* **47** (i), 945. 1914.

(25) WALKER and CORMACK. *Journ. Chem. Soc.* **77**, 5. 1900.

(26) WARBURG, E. *Biochem. Journ.* **16**, 174. 1922.

(27) HASTINGS and SENDRAY. *Journ. Biol. Chem.* **65**, 445. 1925.

(28) GAARDER, T. *Bergens Museums Aarbok*, 1916–1917, Heft 1, 66.

(29) IRVINE, L. "The Carbonic Acid-Carbonate Equilibrium and other Weak Acids in Sea Water." *Journ. Biol. Chem.* **63**, 767–78. 1925.

(30) ATKINS, W. R. G. "The Hydrogen-Ion Concentration of Sea Water in its Biological Relations. Part I." *Journ. Mar. Biol. Assoc.* **12**, 717. 1922. Part II. *Ibid.* **13**, 93. 1923. Part III. *Ibid.* **13**, 437. 1924.

(31) LEGENDRE, R. *La concentration en ions hydrogène de l'eau de mer.* Paris, 1925.

(32) *United States Coast Guard. Bulletin No.* 12. *International Ice Patrol, Season* 1924. Washington, 1924.

(33) ATKINS, W. R. G. "Silica Content of Natural Waters." *Journ. Mar. Biol. Assoc.* **13**, 151–57. 1923. *Ibid.* **14**, 89–99. 1926.

BRANDT, K. "Über den Stoffwechsel im Meere." *Wiss. Meeresuntersuchungen*, **18**, 185–430. Kiel, 1920.

(34) STOWELL, F. P. "The Purification of Sea Water by Storage." *Proc. Zool. Soc. London.* April 1926.

(35) BRANDT, K. "Über des Nitratgehalt des Ozeanwassers und seine biologische Bedeutung." *Abh. der Kaiserl. Leop.-Carol. Deutschen Akad. d. Naturforscher*, **100**, Nr. 4. Halle, 1915.

(36) ROBERTS, C. H. "Oil Pollution." *Journ. du Cons. Internat.* **1**, 245–75. 1926.

(37) ATKINS, W. R. G. "A Quantitative Consideration of some Factors concerned in Plant Growth in Water." *Journ. du Cons. Internat.* **1**, 214–16. 1926.

(38) IRVINE, L. "The Precipitation of Calcium and Magnesium from Sea Water." *Journ. Mar. Biol. Assoc.* **14**, 441–46. 1926.

HAAS, A. R. "The Addition of Alkali to Sea Water." *Journ. Biol. Chem.* **26**, 515. 1916.

(39) SAUNDERS, J. T. "The Hydrogen-Ion Concentration of Natural Waters. I. The Relation of *p*H to the Pressure of Carbon Dioxide." *Brit. Journ. Exp. Biol.* **4**, 46–72. 1926.

(40) FOX, C. J. J. "On the Coefficients of Absorption of the Atmospheric Gases in Distilled Water and Sea Water. Part II. Carbonic Acid." *Publ. de Circonstance*, No. 44. Copenhagen, 1909.

TORNÖE, H. *Chemistry. The Norwegian North Atlantic Expedition of* 1876–1878. Christiania, 1880.

(41) ATKINS, W. R. G. "Respirable Organic Matter in Sea Water." *Journ. Mar. Biol. Assoc.* **12**, 772. 1921.

(42) MCCLENDON, J. F. "On the Changes in the Sea and their Relation to Organisms." *Publ.* 252, *Carnegie Inst. Washington*, 213–58. 1918.

(43) GRAN, H. H. and RUUD, B. "Untersuchungen über die im Meerwasser gelösten Stoffe." *Avhand. utgiff av Det Norske V.-Akad. i Oslo.* I. Matem.-Naturvid. Klasse, No. 6. 1926.

(44) MARSHALL, S. M. and ORR, A. P. "The Relation of the Plankton to some Chemical and Physical Factors in the Clyde Sea Area." *Journ. Mar. Biol. Assoc.* **14**, 837–68. 1927.

Chapter III

WATER MOVEMENTS

The water of the great oceans is not stagnant; a slow interchange of the water masses is brought about by currents. It is only in the deeper water filling the basins of adjoining seas, which are connected with the open ocean over a shallow strait, that stagnant conditions may be approached. There are three distinct types of water movement—tides, tidal currents resulting therefrom and ocean currents—in addition to wave motion brought about by wind.

The *tides* are periodic variations of the water level, due to the attraction of the water particles towards the moon and sun. These changes in water level give rise to regular periodic *tidal streams* which may reach considerable velocity over shoals such as the Grand Banks or in shallow water in the neighbourhood of land. In well-defined channels the stream generally follows their direction, flowing for a period in one direction and then for a similar period in the reverse direction; in the open sea the direction is frequently rotary, setting towards a different point of the compass during each hour of the tide. These tidal streams cause a to-and-fro or oscillating movement of the water, which almost returns to its original position at the end of each period, usually every 12½ hours. *Ocean currents*, on the other hand, may be permanent or seasonal or merely adventitious, and it is through them that the water of the great oceans is in a constant state of circulation. Tidal streams are in many places superimposed upon such currents, in which case the measure of the current is the residual drift after one or more complete periods of the tidal stream.

THE TIDES

A quantitative consideration of the causes of the tides, and of the several theories to which the tides of the world conform fairly closely, is outside the scope of this volume. A practical knowledge of how to find the approximate height of the tide at any stated time from readily accessible data, and the meaning of the terms in common use, is so essential for any work either in the inter-tidal zone or at sea near the coasts that the following information may be of practical utility. The simplified theory of the tides and the forces giving rise to them, included here, fits the facts well enough to be useful as a crude working hypothesis. For more detailed and accurate information of this nature the reader is referred to the various works on tides.

Owing to the variation in gravity occasioned by the attraction of the moon, a wave with two troughs and two crests—one towards and one away from the moon—travels round the Southern Ocean, which encircles the world south of the American, African and Australian continents.

Directly beneath the moon, the water particles are pulled upwards against the pull of the earth owing to the moon's attraction. This only amounts to about one ten-millionth of the pull of earth, and therefore does not move the water particles vertically upwards, but in all positions between this and its antipode the attraction is partly vertical and partly horizontal. The horizontal pull moves the water so that it forms a crest beneath the moon and at the antipode. It is actually a composite wave which runs round the Southern Ocean, for it is caused by the resultant of the attractions of the moon and of the sun on the water particles; the force due to the moon is about $2\frac{1}{4}$ times as great as that due to the sun owing to its being nearer to the earth. Since the relative positions of the moon and sun are continually changing, the resultant force also changes continuously.

This primary wave can be conveniently regarded as giving off derived waves which run northwards up the three great oceans, and the crests of a series of free waves or undulations pass up the Atlantic Ocean from south to north twice a day. While these waves pass over the surface at considerable speed the water itself simply rises and falls with a very slight to-and-fro motion of a rhythmic character; it is the energy and not the mass of water which is being bodily transferred. The progress of these undulations cannot be so regularly traced on the coasts of America as on those of Africa and Europe, where it causes high water to occur at the various places which it passes. In the open deep ocean the height of the undulation from crest to trough does not exceed about 3 feet, and on small isolated oceanic islands the rise and fall of the tide, that is the difference between high and low water level, or *range* of the tide does not exceed a similar amount. As the wave approaches shallowing coasts or submarine banks its height is increased and still further augmented on reaching the coast. The free waves pass rapidly up the Atlantic and reach the open ocean south of Britain in about $1\frac{1}{2}$ days. They can be regarded as dividing into three main lines of advance. One passes up the English Channel and on into the south-east part of the North Sea; this arrives at Dover about the same time as the second offshoot up the Irish Channel arrives at Liverpool, and as the third offshoot passing along the west coast of Ireland round the north of Scotland, arrives at the Shetlands. This third offshoot runs on into the North Sea and passes along the east coast of Scotland and England, running into the Thames estuary 12 hours after passing round the north of Scotland, at about the same time as the next succeeding wave has run up the English Channel and passed through Dover Straits.

The tidal wave running up the English Channel has a greater height on the French than on the English shore, and the wave at the eastern end tends to increase in height where it is constricted as the Channel narrows. In the same manner, owing to constriction, the wave entering the Bristol Channel increases in height to 40 feet at the Bristol end. Part of the wave moving up channel is considered to be reflected from the French coast in the neighbourhood of the Channel Islands, and the reflected wave travelling in a northerly direction causes a double tide on the coast between Portsmouth and Portland (Warburg).

In general, however, two tidal waves occur on the European coast during the lunar day of 24 hours 50 minutes, and usually around our coasts they are nearly equal in height.

In other regions, one of these semidiurnal waves may be slight or even inappreciable, in which case the range of one of the semidiurnal tides is much greater than that of the succeeding tide (Vancouver, B.C.). In the extreme case diurnal tides alone occur (French Indo-China).

As already stated the primary waves passing round the Southern Ocean are composite, being due to the forces exercised by both sun and moon. The combined force approaches its maximum when sun, moon and earth approach the condition of being most nearly in line, that is at full moon, and at new moon or *Change*.

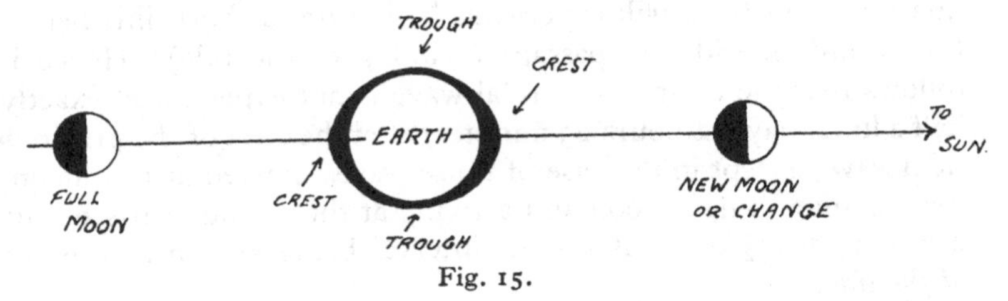

Fig. 15.

On these days the derived wave running up the Atlantic will have the greatest height from crest to trough; from $1\frac{1}{2}$ to $2\frac{1}{2}$ days later this wave will reach the ports round Britain and so give rise every fortnight to tides with maximum range or *Spring Tides* (from Saxon *sprungen*, to bulge).

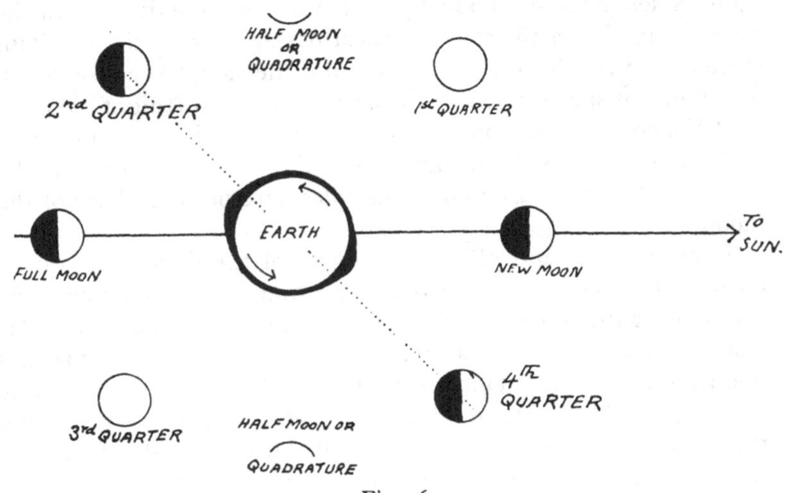

Fig. 16.

At half moon the forces acting on the Primary wave are in opposition, and tides with small range or *Neap Tides* occur.

If we consider the position of the crest of the composite tide when the moon is in its 2nd or 4th quarter, it will clearly lie in advance of the meridian cut by the line moon to centre of earth, and an observer in the Southern Ocean will experience high water a short time after the moon has passed his meridian (lagging of the tide). Conversely when the moon is in the 1st or 3rd quarter an observer will experience high water a short time before the moon's meridional passage (priming of the tide). Hence it follows that the crest of one tidal wave is not experienced exactly half a lunar day (12 hours 25 minutes) after the crest of the previous tidal wave, except in the case of those waves derived at full moon, new moon and half moon and arriving at the various parts of our coasts 1½ to 2½ days later—this interval being known as the *age of the tide*.

So far we have considered the angles made by moon and sun with a meridian plane, that is with a plane running through both poles of the earth.

Because the angles (or declinations) made by moon and sun with the plane running through the equator, that is the plane lying at right angles to the meridional planes, vary, and the distance of the moon from the earth in its elliptical passage varies, the resultant force of the two heavenly bodies is not the same every fortnight. The range of spring tides is not the same every fortnight. Where, as at Vancouver, B.C., one of the two daily tides is small compared with the other it is the declination of the moon which exerts the most effect on the variation in the tides, and not the phase of the moon.

Since the time and place of the crest of the tidal wave, on an ideal earth completely covered with water, depends upon the positions of the moon and sun in three dimensions (their angular distance from a meridional plane, from the equatorial plane and upon their varying distance from the earth), it is evident that any theory of the tides is complicated; if we take into consideration the large and irregular continents and the varying depths of the oceans which affect the rate of progression of the tidal waves, the theory becomes extremely complicated, and complete agreement cannot be expected between theory and observation. In order to predict the tides, observations have been carried out at a number of ports—*standard ports*—extending over a year or more, and from an analysis of these observed data, the actual effect of the relative positions of the moon and sun, of their varying distances from the earth and the deformation of the harmonic form of the combined wave owing to its travel over shallow water can each be calculated. From these factors the tides at any future time and date can be assessed, the astronomical data necessary being known. Such predictions, based upon actual observations at the particular place, are very close approximations.

Tide tables are based on the predictions for these standard ports. The times of high and of low water for every day throughout the year are tabulated, together with the *height of the tide* above (or below) a certain level at high and low water. The level used for these predictions is an arbitrary one, known as the Datum level, and is in most cases the level of low water at ordinary spring tides (L.W.O.S.). In this case big spring tides such as usually occur about the equinox, fall somewhat below datum at low water. This tidal datum is also the level from which the soundings (or depths) shown on many charts are taken.

From these data the *height of the tide* (*i.e.* distance above or below the datum level used) may be found approximately for any time between high and low water at the standard ports by the following simple method, when there are two high waters of similar height daily.

As an example, assume that it is desired to find the height of the tide above datum at the Mumbles (Swansea) at 4.30 p.m. G.M.T. on October 1st, 1926. From the *Admiralty Tide Tables*, Part I, high water on that afternoon is at 2 hours 31 minutes p.m. and low water at 8 hours 47 minutes p.m.; at high water the tide is 21·1 feet above datum level and at low water 8·7 feet.

The tide in falling occupies 6 hours 16 minutes and the *range of the tide* during that particular fall or ebb is 21·1–8·7 = 12·4 feet. At 4.30 the tide has been falling for 1 hour 59 minutes or $\frac{119}{376}$ of the entire period of the fall. It is assumed that the fall corresponds to simple harmonic motion and on this assumption a diagram is drawn as in Fig. 17, the angle 57° being $\frac{119}{376}$ of 180°. The level at 4 hours 30 minutes is thus found to be 18·3 feet above the datum level, the tide having fallen or 'ebbed' 2·8 feet since high water at 2 hours 31 minutes p.m. This method gives substantially correct results, provided that the tidal wave is not distorted, as by reflected waves or by running some distance over shallow water. In the latter case the flood or rise in water level frequently occupies a shorter period than the ebb or fall in level, the tidal wave having become steeper in front of the crest, similar to wind-waves running up a shallowing shelf.

In narrowing estuaries, as already mentioned, the range of the tide tends to increase on passing up the channel, the energy of the wave being cooped up between the converging shores. The speed, at which the crest of each progressive wave passes, is for the most part dependent on the depth. In uniform channels such as

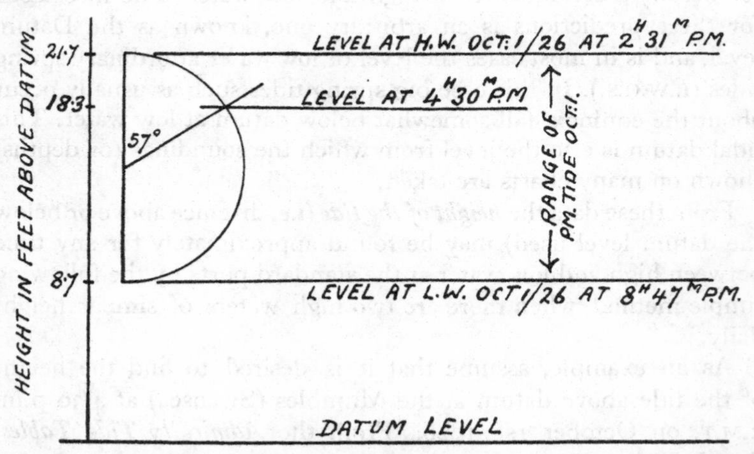

Fig. 17. **Diagram illustrating method of finding the height of the tide between the times of high and low water.**

rivers or canals, on the other hand, the range of the tide decreases. Here the speed of the progressive wave $= \sqrt{gh}$, where g is the force of gravity and h the mean depth.

In addition to the 'standard ports' observations extending for some length of time have been carried out at a large number of 'Secondary Ports.' For these *Tidal Differences* are published, being the difference in time and height of high and low water at both spring and neap tides between the secondary port and the times and heights predicted at a stated standard port. By applying these Tidal Differences in the manner set out in the tide table from which the data are taken, good approximations may be obtained.

For a large number of other ports and positions a certain amount of tidal information is given on charts and in the tables; this generally takes the form of *Tidal Constants*, or the time (local mean time) of high water at full and change of the moon and either the *range of the tide* at springs and neaps, or the *height of the tide* at high water springs and neaps above the Datum level of the chart. In the latter case either the height of mean sea level above the datum or the height of low water springs is also required.

From this information the times and heights of high and low water for any particular day may be calculated roughly, and a rough approximation of the height of the tide at any time may be arrived at. Where there are two tides a day nearly equal in height the results obtained are not likely to be more than an hour out, but where alternate tides differ considerably in height the results obtained by such calculation cannot be relied upon.

As an example of the use of tidal constants consider a position where H.W.F. & C. (high water at full and new moon) = y hours z minutes. This means that when the moon passes the meridian at midnight and at midday high water occurs y hours z minutes later, and when the moon passes the meridian at other times:

Time of high water (local mean time)
= Time of moon's transit + y hours z minutes ± correction for priming and lagging of the tide.

The time of the moon's upper and lower transit for each day of the year is given in almanacks and in some tide tables. This time needs a small correction for the longitude of the place in question, being two minutes earlier for each 15° of east longitude and two minutes later for each 15° of west longitude. This correction converts the time of transit at the meridian of Greenwich which is stated in G.M.T. into the time of transit of the meridian of the particular longitude in terms of local mean time. The correction for the priming and lagging of the tide may be taken from Table XXXIV.

Constants relating to the height of the tide are stated in a number of ways, from which it is usually possible to arrive at:

Height of the water level at H.W. mean spring tides above datum level.
Height of the water level at H.W. mean neap tides above datum level.
Height of mean tide level above datum level.

It is easy to arrive at the interrelation of these with other methods of stating the constants from Fig. 18. If the range of the tide or rise at springs and neaps only is given, an arbitrary datum level may be chosen such as L.W. mean springs.

In some positions the tidal constant for the time of low water (L.W.F. & C.) is included, in which case the times of low water may be calculated. Where no information regarding the times of low water is given it is assumed to occur 6 hours 12 minutes after high water.

Table XXXIV. Priming and lagging of the tide.

Place	Age of tide (days)	Hour of moon's transit 0	1	2	3	4	5	6	7	8	9	10	11	12
		Correction. Minutes												
*	1	0	−16	−31	−46	−55	−58	−42	−12	+18	+30	+26	+16	0
†	1½	0	−16	−32	−47	−58	−64	−57	−27	+ 3	+23	+22	+15	0
‡	2	0	−16	−32	−47	−62	−70	−69	−47	−17	+11	+17	+12	0
§	2½	0	−16	−31	−46	−61	−72	−76	−64	−34	− 2	+12	+10	0

* West coast of Africa south of lat. 30° N.; east coast of America between lat. 40° S. and lat. 50° N.

† East coast of South America south of lat. 40° S.; east coast of North America north of lat. 50° N.; Iceland; west coasts of Africa and Europe from lat. 30° N. to North Cape in Norway, including the English Channel, Ireland, and west coasts of England and Scotland.

‡ East coast of Scotland and England north of lat. 52° N.; north coast of Europe.

§ East coast of England south of lat. 52° N.; coasts of Belgium, Holland, Germany, and Denmark.

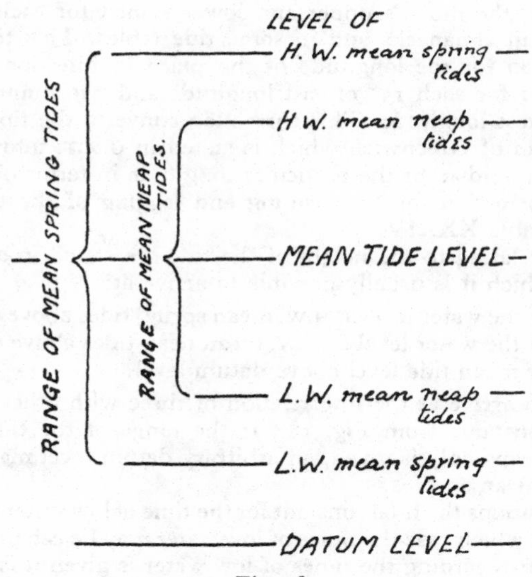

Fig. 18.

Given these three values it is possible to arrive at the height of the tide above datum level at any high and low water, and to calculate the height of the tide above datum at any time by the method shown on p. 89. In the first place it is necessary to know the number of days before or after the nearest spring tide of the particular tide in question. The date of the nearest full or new moon is known and the date of the spring tide occurs x days later, x being the 'age of the tide' at the particular district in question. This value can be obtained from Table XXXIV. With this information a 'Factor' can be obtained from Table XXXV and the heights of high and low water on any day can be calculated from the formulae:

Height of high water = half the sum of the heights of mean high water at springs and at neaps ± half the 'factor' multiplied by the difference in the heights of mean high water at springs and at neaps.

(*Note.* With regard to the sign ±, this is + when the factor is taken out from the table for so many days after spring tide, and − when taken out for so many days before spring tide.)

Table XXXV.

Days from spring tide	0	1	2	3	4	5	6	7
Factor ...	+1·0	+0·9	+0·6	+0·2	−0·2	−0·6	−0·9	−1·0

Having found by this means the height of high water for a particular tide, the height of the previous or subsequent low water will equal the height of mean tide level less the difference between the calculated height of high water and the height of mean tide level.

It is clear that the heights found from these tidal constants are only roughly approximate, since no account is taken of the tidal change due to changes in the moon and sun's declination and distance from the earth, which cause marked changes between one high tide and the next.

In the foregoing the tides have been considered from the point of view of Whewell's progressive wave theory, which has afforded a working hypothesis upon which modern knowledge has been built up. It must be explained, however, that there are cases which cannot be explained on this hypothesis. About 1900 an entirely different theory was put forward by an American mathematician, Harris, which affords a very reasonable explanation of some particular cases where the progressive wave theory appears to break down. Harris considered the tides as stationary waves such as are set up on rocking a trough of water. The water level at each end of the trough continues to oscillate while the level in the middle of the trough remains the same. The period of each uninodal oscillation is $\frac{4L}{\sqrt{gh}}$, where L is half the length of the trough, g the force of gravity and h the depth of the water. The Bay of Fundy is an instance where waves of this nature appear to account for the tides. At the mouth of the bay the tidal range is 13 feet, while at the head of the bay the range is over 40 feet. The shore lines do not rapidly converge nor can shallowing depths account for this on the theory of a progressive wave being cooped up as it passes up the bay. Furthermore, the level commences to rise at the mouth of the bay only 5 minutes before it commences to rise near the head of the bay where the range is 40 feet. The bay, however, is of such length and average depth that a stationary wave would have a period of oscillation of nearly 12 hours, roughly synchronising with the period of the Atlantic tidal wave. The shoal depths at the upper end would also add to the range of a stationary wave.

TIDAL STREAMS

In the open oceans where the depth is great there is practically no horizontal movement of the water due to the tidal wave. Since the height of the wave is only two or three feet, while its length is some hundreds of miles, friction is not overcome. When, however, a tide wave meets a submarine plateau, its height increases considerably, its length diminishes as its speed decreases, and in consequence the gradient from crest to trough becomes sufficiently steep to allow the water to flow from higher to lower level, and such a periodic flow of water is called a tidal stream. Just as the velocity of the stream is increased by shallow water, so its velocity in the neighbourhood of land is increased by shoals, uneven bottom, and salient points of land; strong streams are in consequence frequently experienced off headlands.

Not only does an obstruction increase the rate, but it may also alter the horizontal direction of the stream which meets it, or it may cause considerable turbulent motion among the water particles where the stream runs over an uneven bottom; this causes a 'race' in extreme cases, as off Portland Bill, where the tidal stream runs over a submarine ledge. Even where a tidal stream runs over a shoal, such as the Jones Bank, south-west of the Scillies, which is a hill in the sea bottom about 200 feet high and having 240 feet of water over it, a disturbance of the sea surface is apparent even in the finest weather. In this way tidal streams cause mixing of the bottom water, where nutrient salts are regenerated from dead organisms, with the upper layers.

The tidal stream in well marked channels usually follows the direction of the channel and usually, but not always, runs in one direction for 6¼ hours, then reverses and runs in the opposite direction for an equal time. The times of reversal or of 'slack water' do not generally correspond with the times of high and low water, and are rather later a few miles off shore than close in shore. The tidal stream running up a channel, as up the English Channel towards Dover, is often termed the 'flood'—and running down the 'ebb'—although it may not occur entirely during the period while the *tide* is rising or flooding and falling or ebbing respectively; thus at the moment of writing the tidal stream is running eastward or flooding up channel past Rame Head towards Dover, but the level of the sea's surface is falling, that is, the tide is falling or ebbing. The terms flood and ebb have two quite distinct and different meanings when they are applied respectively to the tide and tidal streams.

The velocity of a to-and-fro tidal stream which runs for roughly six hours either way is usually greatest during the third and fourth hours after slack water, and in many places barely perceptible during the first and fifth hours. The velocity attains its maximum when the range of the tide is greatest (Springs), and least when the range of the tide is least (Neaps).

In places other than well defined channels the tidal stream may vary in nature from complete reversal in direction at slack water to rotary tides where there is no slack water and where the direction is continuously changing. Such rotary tides are usual over isolated banks; they occur over the Grand Banks and Rockall.

Tidal streams with a semi-diurnal period are characteristic of coastal areas and submarine banks, while in the deep open ocean there is some evidence that a diurnal period is more pronounced when streaming does occur. During the drift of the *Fram* in the Polar Sea the heavy pressures of ice displayed a diurnal period, and Werenskiold has observed a far more pronounced diurnal period in the tidal streams at a distance from the Norwegian coast than near it.

The direction of the streams is shown on charts by arrows together with the average velocity attained during the third and fourth hour, that is the average maximum velocity, at springs and at neaps, while the times of reversal or slack water are frequently given as so many hours after and before high water at a neighbouring port. For some positions the average direction and average maximum velocity at springs and at neaps are given for every hour of the complete cycle. The velocities are those observed during calm weather; *a strong wind has a marked effect upon both velocity and direction*, for it tends to produce a surface current. Several useful books showing the direction of the tidal streams around the British Isles for every hour after high water at Dover have been published, some of which indicate the velocity of the stream as well as its direction.

Little is known concerning the velocity of tidal streams below the surface layers. A material change in velocity with increasing depth has been observed in the North Sea, in Dover Straits and off Plymouth.

OCEAN CURRENTS

The ocean currents have been investigated by three main lines of attack: (i) by observing the drift of partly submerged objects which are carried in a surface current; for instance, wrecks or plants or animals peculiar to some known locality, such as Sargasso weed from the S. American shore, (ii) by observing changes in salinity of the water at some one position and by examining the general distribution of salinity both at the surface and at different depths, (iii) directly by the use of current meters. In addition to these observational methods, a consideration of the forces which give rise to currents not only throws much light upon the interchange of the water masses occurring in nature, but has recently led to a method of computing the rate and direction of currents from the distribution of density of the water.

These currents are not limited to the upper layers, but may extend to considerable depths; moreover currents running in different directions may be superimposed one upon another.

The earlier investigations were made by noting the tracks followed by derelict vessels and wreckage, and by computing the

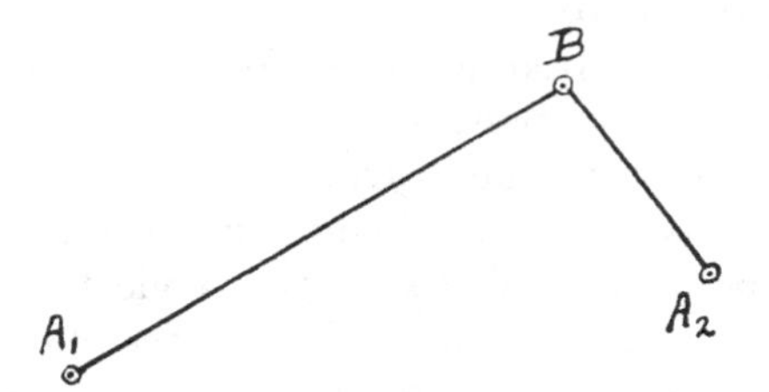

Fig. 19. Method of computing the direction and velocity of a current experienced by a ship.

A_1. Astronomical position of the ship at departure.
A_2. Astronomical position of the ship after sailing x hours.
A_1B. Direction by compass of the ship's track through the water and distance sailed by the ship in x hours.
BA_2. Compass direction and flow of the current in x hours.

distance ships were set off their course by the current in the manner shown in Fig. 19. Bottles and similar objects, weighted so that they have a very slight positive buoyancy and containing a postcard addressed to the investigator, have been sent adrift in great numbers; many of these have been recovered, and from the positions at which they were picked up, the vector average of the surface currents encountered has been calculated. A very ingenious development of this method has been made by Dr G. P. Bidder for investigating the bottom currents in areas where intensive trawl fishing is prosecuted. A yard of copper wire is attached to the neck of a soda-water bottle, and the neck so weighted that the whole sinks, yet remains nearly upright in the water while the tip of the wire trails along the bottom ((1) 1922). In the North Sea, where the method has been used extensively, a fair percentage of these bottles have been caught by trawlers and the date and position of capture notified to the investigator on a postcard provided in

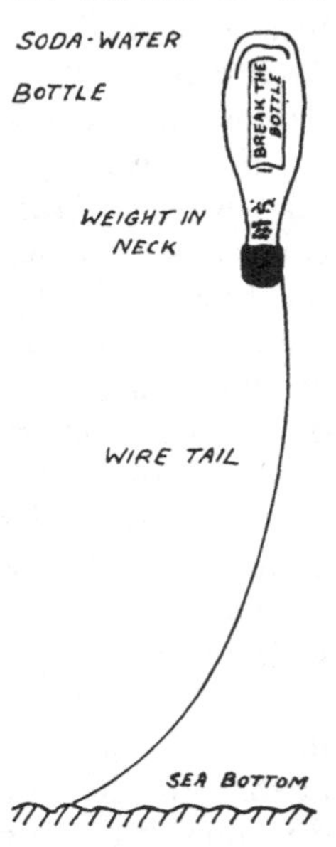

Fig. 20. Bottom trailing bottle poised in the water, resting lightly on its wire tail.

the bottle. A full description of the experiments and a critical review of the information afforded by them has been published by Carruthers ((2) 1926).

The movement of Atlantic water from time to time into the English Channel, and its passage towards the Straits of Dover, was discovered by comparing the distribution of salinity as it occurred every three months for several years. This is a method which has very general application. The necessary collection of samples of water from various depths and the measurement of temperature *in situ* is carried out by means of a metal or ebonite 'bottle' suspended on a wire. It is sent down open at both ends and can be closed when at the required depth by sliding a messenger weight down the wire which hits the trigger of a closing mechanism. For moderate depths an insulated bottle containing a thermometer is used and the temperature at the different water levels can be read after it has been raised to the surface. When water is brought to the surface from great depths, it will naturally be subjected to a considerable decrease in pressure which gives rise to expansion. As it expands it cools, and it follows that when working in great depths the decrease in temperature coupled with the time taken to bring the sample to the surface, will render such thermometer readings inaccurate. A reversing thermometer is used to overcome this difficulty, that is, one in which the thread of mercury breaks when it is tipped upside down. The reversing of the thermometer at the required depth is arranged to occur when a messenger weight sliding down the wire hits the trigger on the 'water-bottle.' The types of instruments commonly used are fully described in the textbooks of Oceanography (3).

In moderate depths and a calm sea current-meters may be employed, either suspended in the water from a moored vessel or from a moored buoy. The most usual type measures the current which flows past, by recording the number of turns a propeller has made due to the flow of water past the instrument; they also indicate the direction of the current at intervals of so many turns of the propeller, the whole apparatus being kept pointing in the direction of the current by a vane. Shot are inserted into a tube leading to a cogwheel geared to the propeller, and at each revolution of this cogwheel one shot is discharged into the centre of a grooved magnetic needle along which it rolls to drop off at its north point, when it falls into one of a number of radiating partitions. The magnet remains north and south, while the instrument including the partitions takes up changes in direction of the current; hence the angle between the partition into which a shot has dropped and the axis of the vane indicates the magnetic direction of the vane and consequently of the current. The meter can be left working for a considerable time—until the supply of shot is exhausted—and on examination the number of shot in each partition shows the number of units of distance the current has flowed in the indicated direction.

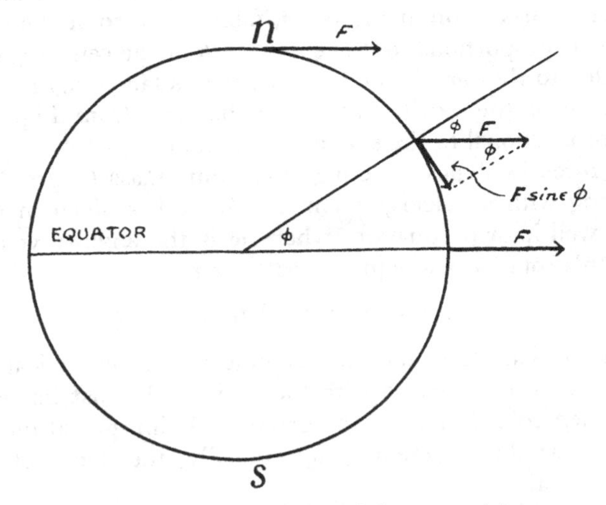

Fig. 21. Diagram illustrating the variation of the tangential component of the centrifugal force (F) due to the earth's rotation, from F at the poles where it is tangential to the sea's surface to zero at the equator where it is normal to the sea's surface. In latitude ϕ it equals F sine ϕ.

Causes which give rise to ocean currents, as distinct from the to-and-fro movements of the tidal streams around coasts, are wind and internal changes causing an alteration in the specific gravity. The latter are brought about by evaporation from the surface, rainfall or the melting of ice, and by changes in temperature. Such changes give rise to *Archimedean forces* in the sea.

Whatever the cause, once a current of water-particles is set in motion it is subjected to a pull (to the right of its direction of movement in the northern and to the left in the southern hemisphere) owing to the rotation of the earth. The deflective force acting upon the particle is:

(1) Directly proportional to its mass.
(2) Directly proportional to its horizontal velocity.
(3) Exactly the same whatever its horizontal direction.
(4) Always at right angles to its direction of movement at the moment and therefore without influence upon its speed.
(5) To the right of the course of the particle in the northern hemisphere and to the left in the southern hemisphere.
(6) At a maximum at the poles, falling to zero at the equator. It is proportional to the component of the centrifugal force, due to the earth's rotation, which lies tangential to the surface of the earth, and as can be seen from Fig. 21 it is proportional to the sine of the latitude.

The force in dynes F acting upon unit mass (1 grm.) which is moving with a velocity v cm. per second is given in the following well-known equation, where ω is the angular velocity of the earth's rotation and equals 0·00007292.

$$F = 2\omega v \text{ sine latitude.}$$

Both this deflection, and the insignificance of the vertical dimension of the sea compared with the horizontal, must be borne in mind when considering ocean currents. A thin pencil line across this page would represent, proportionally, the depth of a deep ocean in section.

Considerable frictional retardation occurs between flowing water particles in contact with the bottom of an ocean basin. Ekman has suggested, as a result of mathematical investigation, that if a steadily proceeding ocean current moves over a gradually shallowing shelf, it will tend to be deflected more and more to the right in the northern hemisphere. If its direction is from shallower to greater depths, then the current will be deflected to the left. It is hardly necessary to add that where a current meets a coast, the configuration of the coast line will affect its direction. Even small differences in depth will influence the stream lines to adapt themselves to the contours of the bottom.

FORCES DERIVED FROM WIND

The wind exerts a tangential force upon the surface of the sea, and sets the particles of water in motion. They are then subjected to a second force due to the earth's rotation, which, acting at right angles to their direction of movement, causes a deviation of this direction from the wind's path—to the right in the northern hemisphere. The layers adjacent below the surface are dragged along by friction, their velocity decreasing with increasing depth, while their direction of motion tends to deviate more and more from that of the wind, unless constrained.

Since water is inextensible it follows that when it moves out from any area as a wind-blown current, other water will flow in from around and from below to take its place, and water ahead of the current will tend to be piled up. The direction and velocity of wind-blown currents are constrained by these compensation and gradient currents; in the oceans they are often further complicated by currents due to Archimedean forces.

The tangential force exerted by the wind on a rough sea has been calculated from the height to which water was piled up against the coast during a gale in the Baltic. If T represents the force in dynes exerted upon one square centimetre of the sea's surface by a wind of velocity W cm. per second, then

$$T = 0{\cdot}0000032\, W^2,$$

a value slightly higher than that found by G. I. Taylor for the tangential force of the wind blowing over the grass of Salisbury Plain.

Interesting and valuable calculations have been made of wind-blown currents under the hypothetical condition where they are not affected by compensation or gradient currents and have attained a steady velocity ((4) 1905, 1923). For a deep ocean under these conditions the surface layer will be deviated 45° to the right of the wind's direction in the northern hemisphere and 45° to the left in the southern. The depth at which increasing deviation will cause a complete reversal in direction of the current has been calculated; this depth is dependent upon the internal friction or viscosity of the water. If the water-particles moved horizontally and had no movement up and down past each other, the viscosity would be that for laminar motion—a known value. In this case the currents would not extend to a depth greater than a few centimetres.

Actually there is movement of the water particles up and down past each other and descending particles give up a part of their momentum to others which they meet, so increasing the velocity of particles at a lower level. In this way turbulence or eddy-motion in the sea causes the water to behave as if its viscosity was often many hundred times greater than its viscosity for laminar motion (20).

Under the hypothetical conditions of no compensation or gradient currents, the horizontal velocity in deep water decreases with increasing depth in geometrical progression, once a state of steady motion has been attained.

$$V_h = V_0 \epsilon^{-\frac{h}{\mu}.\rho\omega \text{ sine latitude}}$$

where V_h is the velocity at depth h, V_0 is the velocity at the surface, ϵ is 2·718.., ρ is the density of the water and μ is the virtual or eddy viscosity.

It is here assumed that μ does not decrease with depth. Its value depends upon the irregular part of wave motion, not upon the regular harmonic motion. This decreases rapidly with depth, but it does not necessarily follow that the irregular motion decreases in the same manner.

There is reason to believe that wind-impelled currents approach a steady velocity within several hours and die away rapidly on the cessation of the impelling force. The gradient and compensation currents set up by them develop and die away more gradually.

Ekman's calculations also indicate that in shoal water the deviation of the current from the wind's path is less, and that when the depth approaches about five fathoms the current flows very nearly in the same direction as the wind at all depths.

These hydrodynamical considerations indicate that *the greater the turbulence, the greater is the depth to which a wind-blown current is able to extend.*

The factors which give rise to turbulence within the water are irregular wave motion, and tidal streams running over an uneven bottom. Vertical movement within the eddies is unrestrained in water of homogeneous density, whereas in water where the density changes with depth it is damped. A method of determining the virtual or eddy viscosity under various conditions is badly needed.

With regard to the deviation of the current from the wind's direction, this is seriously affected by compensation and gradient currents and by the proximity of a coast.

During her famous drift in the Arctic basin, the course of the *Fram* averaged 28° to the right of the wind's direction. Numerous cases in the open oceans have been cited where the wind-driven currents deviate 20° to 40° from the wind's path. In the neighbourhood of a coast both direction and velocity are influenced by gradient and compensation currents to a marked degree. Witting's data of the surface drift in the Baltic show that the current runs nearly twice as fast with a wind blowing up and down the coast as with one blowing directly off shore or directly on shore ((8) 1909).

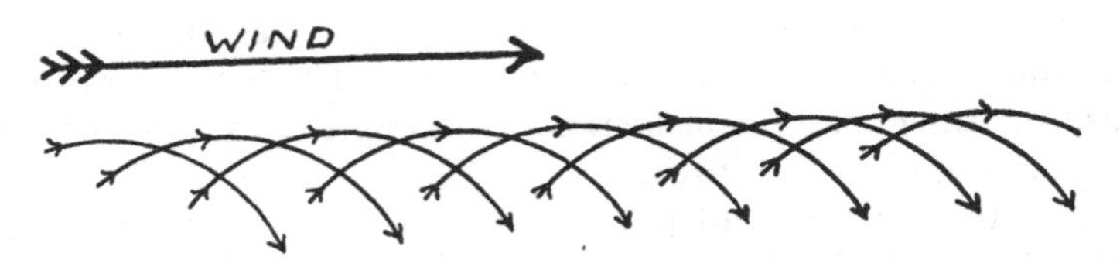

Fig. 22. Diagram illustrating how a wind-driven current may cause little or no translocation of the water.

Drift bottles move up the English Channel into the North Sea during the summer months at an average speed of about 6 miles a day, whereas in the western part of the English Channel no body of water moves eastward at this speed. If it did so water with salinity characteristic of the mouth of the Channel would travel eastward at a notable speed, which it does not appear to do except on rare occasions. It appears probable that the wind-blown surface particles, after travelling a short distance, sink, to be replaced by particles from below, and in this way translocation of water masses with particular characteristics, containing characteristic organisms, does not take place at anything approaching the speed of a floating object (see Fig. 22).

In the English Channel ((5) 1898 and 1927) and in the southern part of the North Sea ((6) 1925) the wind drift takes place in almost the same direction as the average direction of the wind itself.

At the eastern end of the Channel and in the southern part of the North Sea strong tidal streams in the shallower water with uneven bottom set up increased turbulence in the water which is here homogeneous, and in these areas wind-blown currents extend probably deeper and appear more effective in transporting planktonic organisms.

When the effect of a prevalent wind is superimposed on the Archimedean forces tending to set up a current, then considerable bodies of water may be carried in the resulting current. Such a combination is presented by the west going Gulf Stream, which is driven by the north-east Trade Winds and at the same time is acted upon by forces, arising from differences in density, which will tend to give it a westward movement (see p. 112). Here an enormous body of water moves forward along the surface of the ocean towards the north coast of South America with a velocity of some 24 miles a day, which gradually decreases with depth until at about 150 metres relatively still water is reached. It is clear that the friction of the prevalent north-east Trade Wind must aid this movement very considerably, although it is not the preponderating cause ((20) 1918).

Both the simple wind propelled currents in deep water, and currents such as the above, where wind aids the Archimedean forces, are in the customary sense ocean currents, for both carry with them objects floating at the surface; yet their effects on neighbouring regions will be very different. Whereas the first only transports water-particles for a limited distance while they are in the upper layers, the second will move large bodies of water with definite chemical and physical characteristics, as well as their specific flora and fauna, into new regions.

The gradient currents, caused by a wind-blown current piling up water before it or sucking water in behind it, may move considerable bodies of water. Where such a current impinges upon a shore the water level is raised and Archimedean forces are set up which cause a downward streaming, the sinking water being replaced by surface water running in from seaward. Wind blowing on shore tends to raise the water level and off shore *vice versa*. An on-shore wind will set up a current which can be readily demonstrated (Fig. 23). Such winds, driving particles of water against boundary surfaces and giving rise to a gradient, are among the most important of the *external* forces assisting to maintain the circulation in the oceans.

Schott gives two striking instances of the effect of off-shore and on-shore winds ((7) 1924). The surface temperature of the sea on the windward and leeward sides of the Galapagos Islands differs by 11° to 12° (Fig. 24). On the windward side warm surface water is heaped up and on the leeward shore cold water upwelling from below takes the place of the warm surface water which is blown away as a surface wind drift.

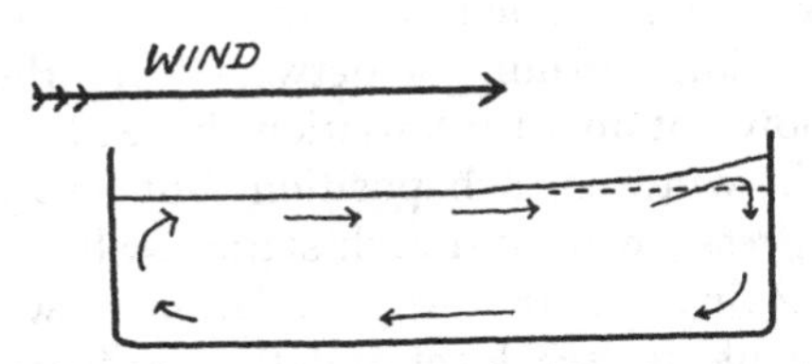

Fig. 23. Experiment illustrating the action of off-shore and on-shore winds upon the movement of homogeneous water.

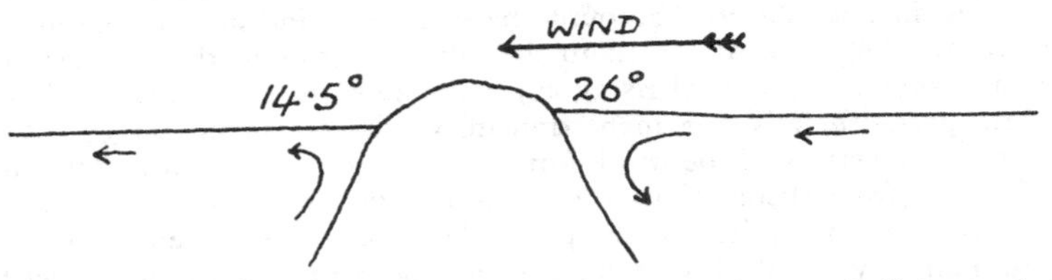

Fig. 24. Diagram showing how the currents due to on-shore and off-shore winds affect the surface temperature on the windward and leeward sides respectively of the Galapagos Islands.

The effect of an off-shore wind on the sea temperature at a Baltic bathing resort in June 1889 was to lower the water temperature from 23° to 6° in a few hours; Krummel quotes a case where, in August in the Memel Deep, after long continued east winds, the coastal temperature fell from 19° to 8° in the course of a day for a similar reason, while it was still 18° at four or five miles off-shore.

The prevalent winds on the west coast of America between the parallels 20° and 45° N. blow the surface water out to sea and this is replaced by colder water from below. From the lowering of the surface temperature due to this, and also from changes in salinity, McEwen has calculated that upwelling takes place at a velocity of about 30 metres per month.

Around our coasts the height of the tide at high and low water, as well as the time at which these occur, is materially affected by strong winds and variations in barometric pressure. On the tideless coast of Egypt a strong on-shore wind will raise the water level some two feet. In the Baltic, variations in the height of the sea level have been the object of extensive investigations by Witting ((8) 1918).

From observations carried out over several years at a large number of positions forming a network over the Baltic, it was possible to follow not only the variations in sea level, in barometric pressure and in wind at each position, but also the distribution of barometric pressure around each station, which gave rise to the winds in the locality. In this way a relation between the gradient or deformation of the sea level and the gradient or distribution of air pressure was arrived at.

The heights of the tide predicted from the forces exerted by the moon and sun do not take into account the effect of wind and of barometric pressure. The sea with regard to the latter behaves in the manner of a water barometer, the level rising approximately one foot when the barometric pressure falls one inch, transitory currents being set up. With regard to the effect of the wind, formulae have been evolved by Ortt and by Witting, from observations in Holland and the Baltic respectively, which agree with observed facts at the particular places where the observations were made. Valuable investigations have been made on this subject by the Liverpool Tidal Institute. When the movement of this wind drift is constrained by an irregular coastline, as around England and Scotland, a directly on-shore wind does not necessarily have the maximum effect; thus on the east coast of Scotland it is a westerly wind which is most effective, presumably owing to an indraught of water from the Atlantic into the North Sea ((9) 1923). Wind is most effective in its action on water which is shallow. The effect of wind and air pressure on sea level has been investigated frequently; it is of economic importance, for many of the largest ports are situated in comparatively shallow water, and navigation of the large ships that are now in common use can only be carried on when the tide is sufficiently high. For the same reason it is not infrequently necessary to load part of the cargo from lighters in deeper water than at the quayside, and in this case an accurate prediction of the influence of the wind on the depth of high water would permit the ship to be loaded to a narrower margin and save part of the expense of loading from lighters.

Besides the simple case of downward streaming brought about by on-shore winds, more complicated movements may be occasioned under particular circumstances. Thus when the water in a closed basin is in layers of definitely unequal density, the action of a wind will set up a streaming movement as depicted in Fig. 25 *a*. As shown later, mixing of the heavier and lighter water across the discontinuity layer does not take place readily; consequently the upper layer is set into circulation by the wind while neighbouring particles in the lower layer are carried along by friction. Such an effect has been known to arise in the Baltic (Fig. 25 *b*).

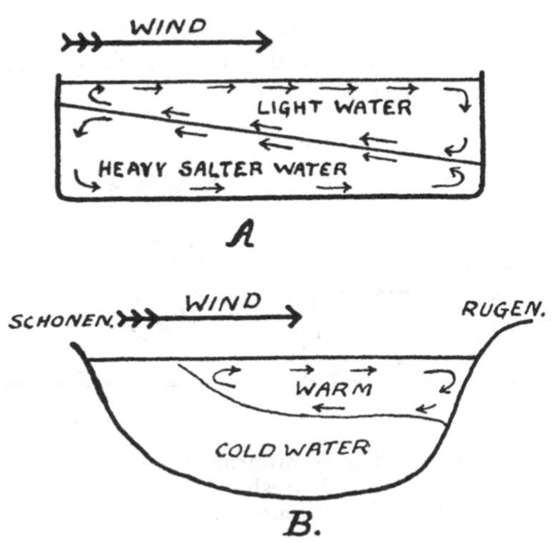

Fig. 25. *A*. Experiment illustrating the action of on-shore and off-shore wind on water in strata. *B*. Wind and distribution of temperature with probable movement of the water in the Southern Baltic on August 1, 1907 (Sandström).

The increase in specific gravity with depth is of frequent occurrence in the sea; in some places the increase is very sudden at a particular depth. A marked 'Discontinuity' layer results, where several surfaces of equal density lie crowded close upon one another. In some cases the density increases more gradually with the depth and surfaces of equal density are more evenly distributed in all depths of the water. If we imagine a water sample from one of these surfaces in the sea transferred to a greater depth, it will be lighter than its surroundings and therefore rise again to its original position. Similarly if it is transferred to a lesser depth it will sink until it reaches the surface corresponding to its own density. It is otherwise, however, when a sample of water is moved in a layer, that is, along a surface of its own density, so that its own density remains equal to its surroundings and no force arises which would occasion its return to its original position. Thus water will move along the density surfaces rather than transversely through them. Moreover eddy motion is restrained where the density surfaces lie close together, the virtual viscosity of the water is less, and consequently less force is required to slide the layers over each other. The forces required to effect vertical translation when the density surfaces are close together is very considerable. (See p. 141.)

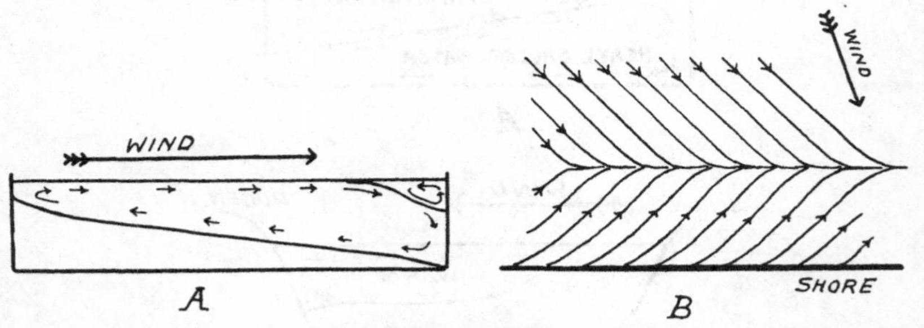

Fig. 26. *A*. Diagram illustrating experiment showing current opposed to the wind on a lee shore against which fresher and lighter water is banked up. *B*. Diagram showing in plan the currents set up close to the shore in Gulmar Fiord.

It is with differences in specific gravity or density of samples at the same depth that we are concerned. Since these differences approximate to those between the same samples at atmospheric pressure, the value at atmospheric pressure is referred to. Formulae and tables for the compressibility of sea water have been constructed by Ekman (10).

However, the vertical equilibrium or stability of the layers is modified by the water's property of compressibility where the increase in density with depth is gradual. Consider a water-particle transferred to a greater depth. Its density will be increased owing to compression, and also owing to compression heat will be generated and its temperature raised, whereby its density is lowered. Hence in arriving at a value expressing the vertical stability of a column of water, it is necessary to allow for the adiabatic heating due to compression of the sample, which is considered as being transferred from a lesser to a greater depth ((11) 1914).

Another type of movement brought about by off-shore winds occurs where a fringe of relatively light fresh water, from rivers, etc., lies along a lee shore, when a circulation may be set up as shown in Fig. 26. The incoming heavy saline surface drift sinks below the less saline shore water, setting up a reversed current in the body of fresher water as described above. This occurs in Gulmar Fiord and generally along the west coast of Sweden (Fig. 26). The line where the currents of salt and less saline water meet is often marked by froth and drifting weed carried respectively from seaward and from shoreward by the opposing currents.

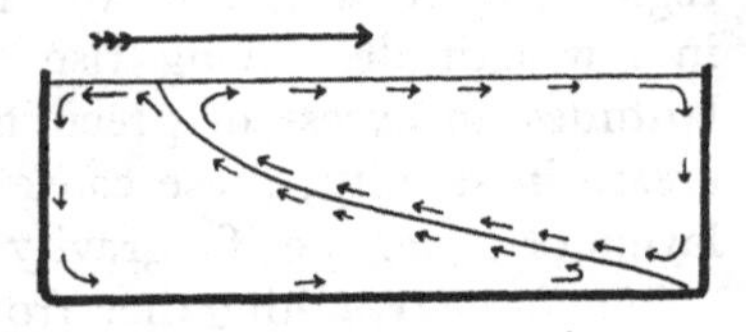

Fig. 27.

Where well-marked layering with a sharp change of specific gravity occurs between one layer and the next, the force required to mix one layer with another is so great that a condition may be set up as shown in Fig. 27. Here the deeper water is brought to the surface on the windward shore and its circulation may even be in a direction opposed to the direction of the wind.

In the open ocean well away from land, where a regular anticyclonic wind sets up a current running in a clockwise direction, such as the Sargasso current, the lighter surface water is deflected to the right into the centre, in the northern hemisphere, owing to the earth's rotation, carrying with it floating weed and other objects (Fig. 28). Thus the warmer lighter water tends to form a pool deepest in the centre (Fig. 29). Such a pool-making effect has been demonstrated by arranging an anticyclonic whirl of air to blow on the surface of water in a tank containing two layers, the upper layer being fresh water coloured with dye and the lower layer salt solution. A regular anticyclonic wind has the effect of driving the lighter water inwards to form a pool, while a cyclonic wind distributes the surface water outwards ((12) 1913).

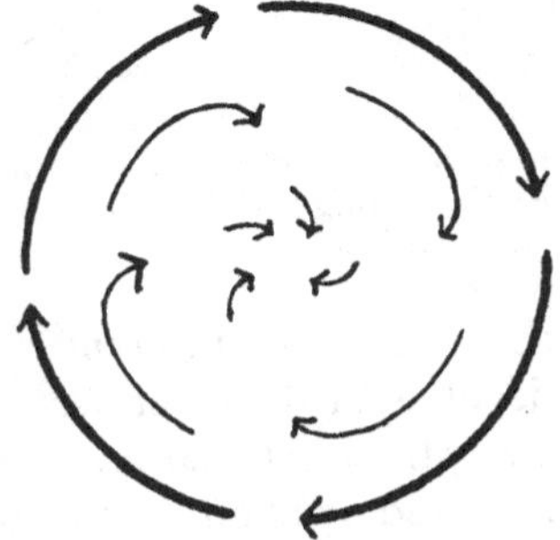

Fig. 28. Anticyclonic wind and current set up by it converging inwards towards the centre in the northern hemisphere.

ARCHIMEDEAN FORCES IN THE SEA

Besides the difference in temperature in equatorial and arctic regions, there is in the North Atlantic an excess of evaporation in low latitudes giving rise to increased salinity, and in high latitudes an excess of precipitation of moisture producing a decrease in salinity; these cause a continuous redistribution of the layers of equal specific gravity.

The forces resulting therefrom, often modified by external forces due to wind, set in action ocean currents which are not restricted to the single upper layer which has become thinner at one end and thicker at the other owing to evaporation and precipitation. It is upon these currents that the task of bringing about interchange between

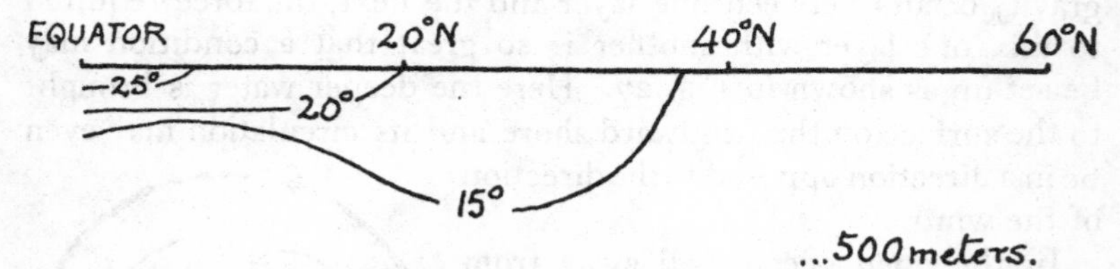

Fig. 29. Temperature distribution in a vertical section through the North Atlantic, long. 30° W., after Schott, showing the 'pool' of warm water in the Sargasso Sea.

the waters of different regions and different depths appears to devolve for the most part. Although impossible to assign quantitatively the modification or reinforcement of the Archimedean forces by wind friction in the open ocean away from land, merely to envisage the violence of the forces at play during a gale leads one to conclude that their effect can be no slight one in those regions where high winds, mostly in one direction, are frequent. The following extract from the narrative of a voyage across the Pacific portrays very vividly the driving force upon the surface of the ocean during a period of violent winds. "Dawn broke on a wild scene. Great white crested seas came sweeping up, one after the other, full of menace, and the air was thick with spindrift blown off the tops of the waves. The pressure of the wind was extraordinary. It howled in the rigging and, in the squalls which followed each other at shorter and shorter intervals, the sound rose to a scream of diabolical rage. As each sea approached it seemed as if it must come bodily on board and sweep everything off the decks." The wind clearly plays a predominant part in governing ocean currents which cause the drift of partly submerged objects—the customary sense of the term 'current'—but the biologist is most concerned with the currents that cause an interchange of considerable water masses, and it is these which have been the main study of the Scandinavian hydrographers. A surface wind drift in another direction may be superimposed upon such currents in some cases.

The primary cause of ocean currents is by no means agreed upon. Some have ascribed it almost wholly to Archimedean forces and others to wind friction. Whichever be the predominating cause in the first instance, a consideration of the types of currents set up by internal forces, irrespective of the external forces due to wind, throws light upon the circulation which actually occurs in the oceans.

When a localised area of surface water increases in density, owing to cooling or evaporation, and becomes heavier than the water below, it falls, lighter water from the surface around taking its place. Owing to the earth's rotation a cyclonic current is formed around the sinking centre in the northern hemisphere in an open ocean.

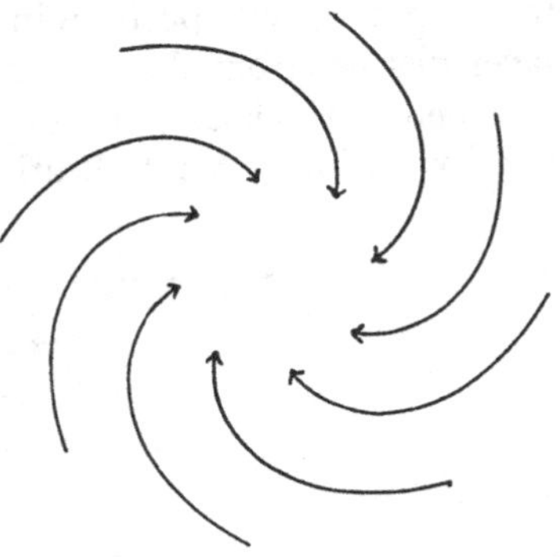

Fig. 30. Cyclonic current formed around a sinking centre in the northern hemisphere.

Such cyclonic currents are a feature of Arctic conditions where the warmer saline water, transported from lower latitudes in the Atlantic, sinks, having cooled in the surface layer and become denser than the colder less saline water below. The heavy water tends to fall in vortex rings, and at the same time considerable mixing takes place. This may be readily demonstrated by allowing ink to drop into a cylinder of fresh water.

In the partially enclosed Mediterranean, evaporation during the summer months causes an increase in salinity and hence in specific gravity, and this warm saline water pours out through Gibraltar Straits as a subsurface current, less heavy Atlantic water flowing in on the surface to take its place (Fig. 31). The warm saline water

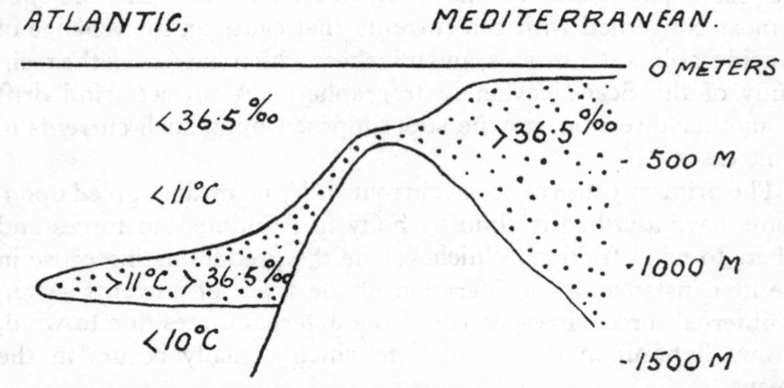

Fig. 31. Diagram showing warm saline water flowing from the Mediterranean through Gibraltar Straits into the Atlantic and forming an intermediate stratum. (After J. Schmidt, "Oceanography of the Gibraltar Region," *Nature*, **109**, 45, 1922.)

takes up its gravity position in the Atlantic forming an intermediate layer distinct from the colder ocean water above and below.

When a localised area of surface water decreases in specific gravity, owing either to heating or to dilution, it tends to run out

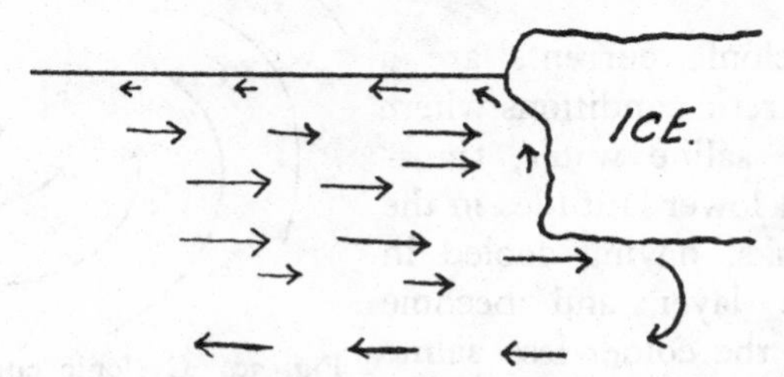

Fig. 32. Currents set up by the melting of ice.

in all directions as an anticyclonic current in the northern hemisphere, unless constrained by land or by wind. The north equatorial or Gulf Stream water in the tropics driven westward by the north-east trade wind and earth rotation is an example.

Very marked currents are produced by the melting of ice in salt water ((13) 1904). In the North Atlantic the 'Arctic Water' of low salinity derived from melting ice flows southward as the Labrador current, hugging the right-hand shore. The deflection to the right owing to the earth's rotation causes it to run into the bays on the Canadian coast. "On the south-east coast, so many wrecks have occurred, especially near Cape Pine and St Shot's Cove, that the compass has been considered to be subject here to local disturbance, but special examination has shown that this is not the case and that these disasters were mainly attributable to the effect of the currents." (Sailing Directions for the N. American Coast.)

Where cold, less saline Arctic water meets and mixes with warmer and more salt water of similar specific gravity, the mixed waters will be heavier than either component, since the salt water has the lesser specific heat.

Table XXXVI.

Salinity, parts per thousand	Specific heat
0	1·000
15	0·958
25	0·945
35	0·931

The mixed water falls, a phenomenon known as *cabelling*.

South of Newfoundland cold water of low salinity, derived in part from the Arctic and from rivers, and covering the slopes of the Grand Banks, meets the highly saline Gulf Stream water. A sharp line of demarcation results and a '*cold wall*' occurs between the Gulf Stream and colder mixed waters. The bow of a ship may be in cold blue-green water at 1° C. while the stern is in warm blue Gulf Stream water at 13–14° C., with its salinity very little reduced by mixing with the colder water. In such positions the isotherms and isohalines are often nearly vertical. The east running Gulf Stream water moves with a velocity in this area of 10 or more miles per day, and the cold mixed waters within a few miles of the 'wall' run parallel with it. This shallow ocean area south of Newfoundland and south and west of Nova Scotia, with depths of less than 100 fathoms, is of particular interest and of great economic importance, since one of the richest and most extensive fishing grounds known exists in this region.

CURRENTS IN THE NORTH ATLANTIC

The usual charts showing the ocean currents relate to the surface two or three metres and are very largely based on records of the drift experienced by ships. The currents as depicted are partly due to the surface wind drift, and in some areas are probably due almost entirely to this type of movement, which affects the course made good by ships and the drift of floating objects but does not necessarily transport any considerable body of water. Such charts based on actual observations have been constructed for each month of the year, the difference from month to month being due in part to seasonal variations in strength and direction of the main currents and in part to seasonal variation in strength and direction of the prevalent winds causing variations in the wind drift.

An attempt has been made in Fig. 33 to compile, from various sources, a rough chart of the North Atlantic showing the currents which carry with them considerable bodies of water, and through which the circulation of the ocean waters takes place, mainly owing to the effect of Archimedean forces. These currents or drifts extend in some cases to considerable depths, such as 1000 metres or more, but at a depth of over 2000 metres it appears from the available evidence that they become almost insensibly slow. Taking into consideration the great depth which amounts to more than 4000 metres over the greater part of the ocean, this subsurface current system is in reality limited to the upper strata, more particularly in the open away from land and submarine banks.

The continuous processes entailing loss of water in the tropical regions from evaporation and gain in the Arctic from precipitation are only in part compensated by the southward flow of water in the Labrador current; there exists in addition to this system of currents a slow drift of cold less saline water in the depths of the ocean. From the Arctic and Antarctic where the incoming water is cooled below the ice, and falls, it slowly streams as deep-seated currents to rise again finally in the equatorial regions where it takes the place of water lost by evaporation and of water carried away in the Gulf Stream.

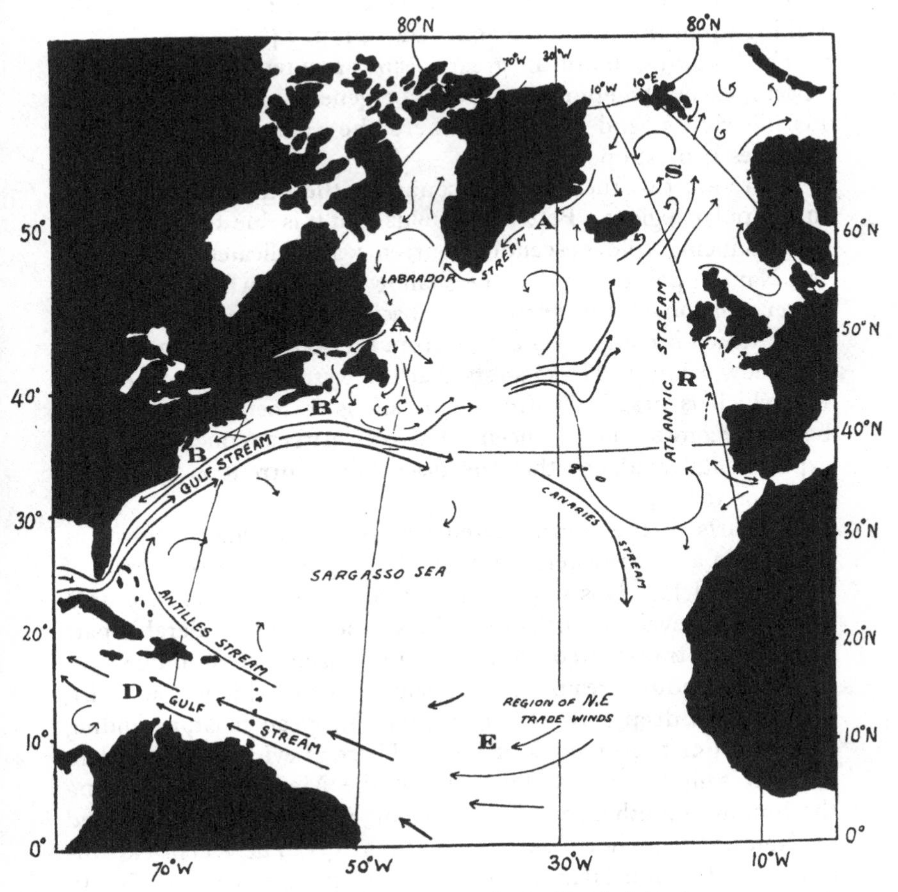

Fig. 33. Subsurface currents in the North Atlantic.

A. South-going light Arctic water hugging the American shore and falling as it mixes with the warm Gulf Stream.
B. Mixture of Arctic, river and Gulf Stream water.
S. Cyclonic currents about falling water.
R. North-going drift of ocean water hugging European shore and replacing south-going Arctic water.
E. Anticyclonic current, anticyclonic wind and rising water.

Recent publications by Merz and Brennecke ((19) 1925) indicate the course taken by the water falling in the polar regions; that originating from the Arctic passes into southern latitudes, rises and turns back running northward between 200 and 1000 metres, finally welling up in the equatorial zone (Fig. 49, p. 146).

The bottom water of the great oceans has a temperature of 3° C. or less, except where cut off from the general circulation by submarine ridges, and is lowest where the connection with polar latitudes is most open.

There are two theories to account for the descending currents in the polar regions. Pettersson holds that it is due to the melting ice producing cold descending currents as indicated in Fig. 32, but Nansen points out that the shallow pack ice which covers the larger portion of the north polar basin floats in a layer of cold water which is completely cut off from the water below on account of its low salinity. He suggests that the water descends when the ice, which is practically free from salt, is formed and the water below becomes more concentrated (3 c). Aitken concludes from experimental evidence that the descending current is cooled by radiation ((13) 1913).

In charts of the ocean currents a stream is usually shown as running in a north easterly direction impinging upon the coasts of the British Isles. It is suggested that this is largely a surface drift before the prevalent south-west winds, whereas a considerable part of the water transported northward, to replace that running south in the Labrador stream and the south-going cold bottom drift, forms a slow deep current hugging the European coast, extending to a depth of 1000 metres or more. There is evidence that warm strongly saline water, welling out of the Mediterranean during the summer months, enters this current; it has been distinguished by Nansen as far north as Ireland ((14) 1913). The recent expeditions of Helland-Hansen and Nansen in the eastern North Atlantic ((18) 1926) yield evidence that the area to the south-west of the British Isles is one of slow cyclonic eddy currents, as deduced from the distribution of density of the water by the application of Bjerknes' Theory; if a considerable north-east going current, such as is often found in the surface layers of the sea to the south-west of Ireland, extended to a considerable depth, evidence of it should be apparent in the density distribution. If, as seems more probable, it is a surface wind blown current, it need not necessarily transport the upper layers for any considerable distance, as the upper water-particles will gradually fall, to be replaced by water from below.

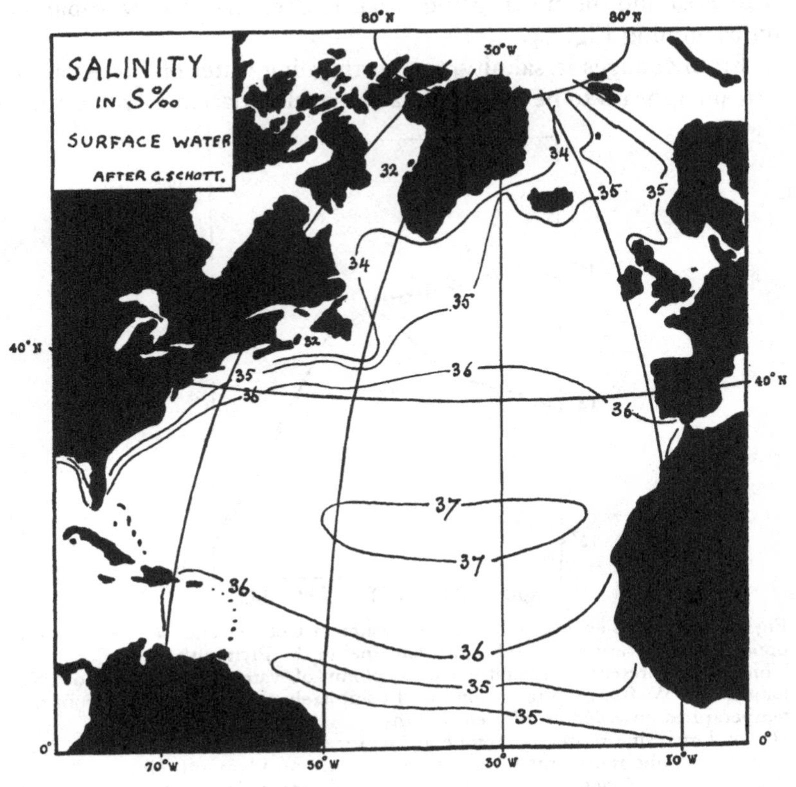

Fig. 34.

The north-going current, known as the Atlantic Stream, flows into the Norwegian Sea over the ridge extending from Iceland to Scotland. A part of it passes round the North Cape into the Barents Sea, where the deep water attains its maximum temperature and salinity in the autumn, after the surface layer has begun to cool. Another part flows into the North Sea, turning north again and running along the Norwegian coast as a strong current.

The effect of this circulation upon the distribution of salinity, at the surface—the pooling of water concentrated by evaporation in the Sargasso, the renovation of the concentrated surface water near the equator, the dilution by the Arctic water in the north-west area and the north-going Atlantic Stream—is very apparent on examining Fig. 34.

From changes in salinity, the north-going water of the Atlantic Stream appears to be subject to a seasonal variation in velocity.

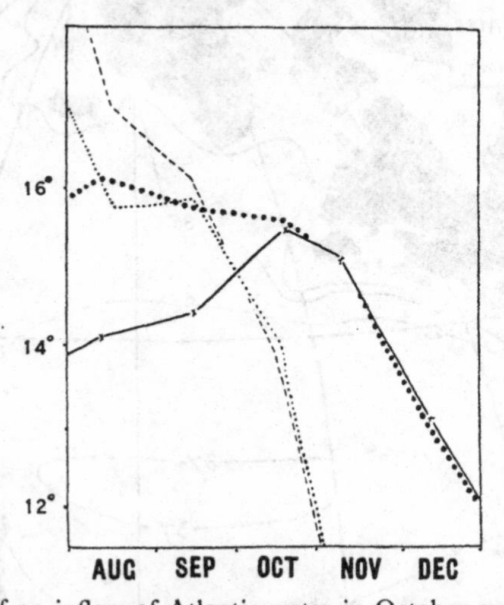

Fig. 35. Effect of an inflow of Atlantic water in October and November 1921 upon the temperature of the English Channel in the Plymouth area.
Thick line represents temperatures of a column of water at E_1, lat. 50° 02′ N., long. 4° 22′ W. from top to bottom, as if completely mixed, calculated from the temperatures observed at different depths.
Heavy dotted line represents temperature of surface water at E_1.
Fine dotted line represents mean air temperature on Plymouth Hoe.
Pecked line represents mean ground temperature 1 foot below surface on Plymouth Hoe.

The mean sea level, after allowing for barometric pressure, is at a maximum at Newlyn in Cornwall and in the North Sea every autumn. This is apparently connected with a seasonal variation in the oceanic circulation. (D'Arcy Thompson, *Fish. Scot. Sci. Invest.* 1914, **4**; Close, *Geographical Journal*, July, 1918; Meissner, *Ann. der Hydrog. und Maritimen Meteorologie*, 291, 1925.)

There is little doubt that it is also subject to fluctuations from year to year; in 1905 and in 1921 the northerly flow was particularly marked, carrying water of high salinity into the North Sea and English Channel. In the latter year the mean temperature of the whole column of water from top to bottom at a position in the English Channel continued to rise after the surface had started to fall in temperature; that is to say, the water in the area was gaining heat owing to the inflow of warmer water while heat was being lost from the surface ((15) 1925).

It is highly probable that the Gulf Stream also shows annual fluctuations which may be of great climatological significance for the countries on both sides of the Atlantic, as pointed out by Helland-Hansen. The *Challenger* and the *Michael Sars*, in 1873 and 1910 respectively, made a series of temperature observations at various depths at nearly the same position in the North Atlantic (37° N., 48° W.). Both found the same temperature at 1000 fathoms and below, but between 1000 fathoms and the surface layers the temperature distribution was very different on the two dates, the differences amounting to nearly 5° C. at a depth of 400 fathoms ((16) 1912).

There are a few instances in which invasions of water from other latitudes into particular areas, resulting from fluctuations in the ocean currents, have brought with them planktonic organisms in quantity which are not ordinarily found in the particular area. Bowman records an unusual incursion of Atlantic water into the Scottish area of the North Sea in 1920–1921 "bringing with it a whole series of organisms not usually found in the North Sea," of which the most striking example was the tunicate *Salpa fusiformis*, and Hardy records an invasion of the pteropod *Limacina*, typical of more southerly latitudes, off the east coast of England in 1921–1922. Gough records in 1904 an unusually dense invasion of the siphonophore *Muggiaea* extending past Ushant into the English Channel as far as Portland, with a branch running north past the Scilly Isles and then turning west and passing round the south coast of Ireland as far as Galway Bay. This animal is typical of more southerly latitudes and usually occurs only in small numbers in this vicinity. Species of *Arnoglossus*, indigenous to more southerly waters, were found in the North Sea during the same year, and an invasion of Salps during the following year (17).

It is necessary to exercise the utmost care in drawing conclusions concerning ocean currents from the composition of their plankton, and it must not by any means be taken for granted that areas where the same species occur are united by a continuous stream connection. Most plankton species are found scattered here and there outside their proper domain, so that these stray individuals might easily become abundant whenever conditions of existence become favourable. It is very difficult to tell what changes in the plankton are due to the direct influence of ocean currents, and what changes are the result of altered conditions partly due to ocean currents and partly to other causes.

It has frequently been observed that the plankton changes its character at the boundary between two currents, as between the Gulf Stream water and the colder water moving south along the United States coast.

Pulses in the currents, which bring deep water from the ocean, rich in phosphate and nitrate, to the upper layers, will carry in train with them not only a slightly different plankton, but modified temperature conditions affecting all marine life. They will also bring nutrient salts which must of necessity in the sunny months affect the phytoplankton species usual in the area, and the marine animals, mostly copepods, which feed directly upon the phytoplankton.

If we consider any particular fertile area of the sea having a regular seasonal variation in sunshine and other meteorological conditions and where the currents undergo a regular seasonal variation, it is not unreasonable to surmise that fluctuations occurring in the fauna are primarily brought about either by fluctuations in the currents or by fluctuations in the meteorological conditions. These would be potent by reason of their altering the regular seasonal round of temperature, available nutrient salts and sunshine for photosynthesis.

SUBMARINE WAVES AND SEICHES

In the open ocean the wind is generally irregular—a series of storms distributed over a large area. Such storms acting upon water in layers of increasing density produce large submarine waves which move slowly forward in the direction of the storm. As the submarine wave moves forward the surface water in front of it must pass over behind the crest of the wave. A strong current against the direction of the wind thus arises in the surface layer.

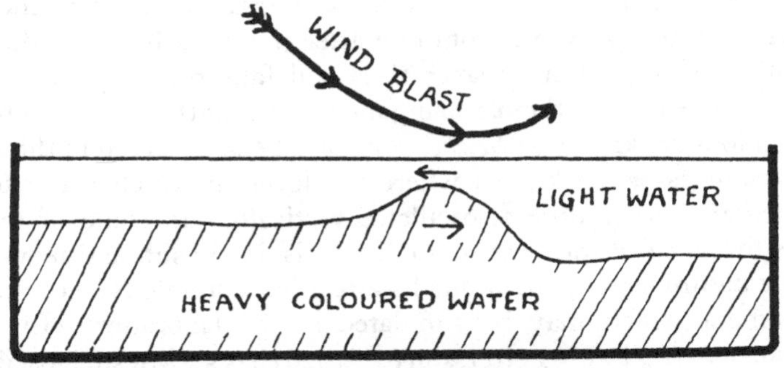

Fig. 36. Experiment illustrating the formation of a submarine wave by a localised wind. (Sandström, *Ann. der Hydrog. und Maritimen Met.* Jan. 1908.)

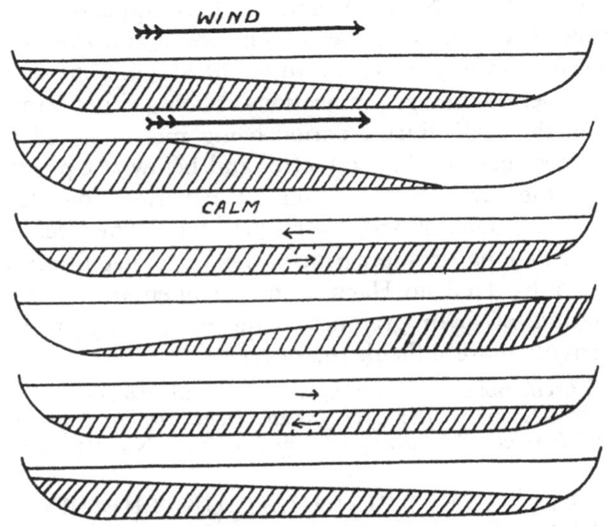

Fig. 37. Diagram illustrating submarine seiches caused by brief gales on stratified water. (Sandström, *Hydrodynamics of Canadian Atlantic Waters*, Ottawa, 1918.)

Such surface-layer currents against the wind have frequently been observed off the Canadian Atlantic coast, where the difference in specific gravity between the layers is often very marked.

The displacement of a marked discontinuity layer by wind may set up a series of oscillations due to the water masses trying to regain their equilibrium as horizontal strata. In consequence a series of stationary waves continue after the wind has died down, so that the discontinuity layer rises and falls regularly for some considerable time, without causing movements at the surface. These oscillations are known as *Seiches*, and have been observed in the sea.

Certain cases are known where the level, at which water of a particular density occurs, oscillates vertically. In some of these cases the cause of the vertical oscillation is due to submarine waves of the nature already mentioned—a relatively simple type of water movement which may be simulated in the laboratory. In the entrance to the Baltic Pettersson ((21) 1926) has shown such oscillations to be tidal waves. Helland-Hansen and Nansen ((18) 1926) have collected evidence of vertical oscillations occurring in various parts of the eastern North Atlantic. There appears to be some connection with tidal phenomena. The phase and the periodicity of the oscillation appear to be different at different depths; in general the periods showed indications of being either semi-diurnal or diurnal, the semi-diurnal period being more marked near the coast and near banks. At 800 metres and below at a position west of Lisbon the level of the water layer varied considerably in depth, to the extent of 200 metres or more. The occurrence of vertical oscillations or submarine waves of a magnitude such as experienced by Helland-Hansen in the open ocean makes the interpretation of hydrographical data of density in terms of ocean currents more difficult than ever.

Vertical oscillations at depths between 75 and 250 metres have been recorded by Defaut in the open ocean (*Deut. Atlant. Exped. auf dem Meteor, IV Bericht. Zeit. der Gesell. für Erdkunde*, No. 5/7, Berlin 1927).

BJERKNES' CIRCULATION THEORY AND ITS APPLICATION

A ship cannot keep a stationary position on the high seas, consequently the use of current meters is strictly limited, owing to error arising from the movements of the vessel herself. Hence any method of attacking the problem in addition to actual measurement of the drift of floating bodies and the study of salinity distribution is an asset of considerable value in studying ocean currents. The application of Bjerknes' circulation theory has proved valuable for this reason.

The physical processes involved in the application of the Circulation Theory may be simply illustrated in the following manner. If a layer of oil moves forward over a stationary bottom stratum of water in a hypothetical tank, which is indefinitely long, the lighter oil will tend to curve to the right in the northern hemisphere owing to the rotation of the earth and the oil layer become thicker on the right-hand side than on the left-hand side of the tank; the separating line will lie obliquely. The obliquity will depend upon (i) the horizontal component of the strength

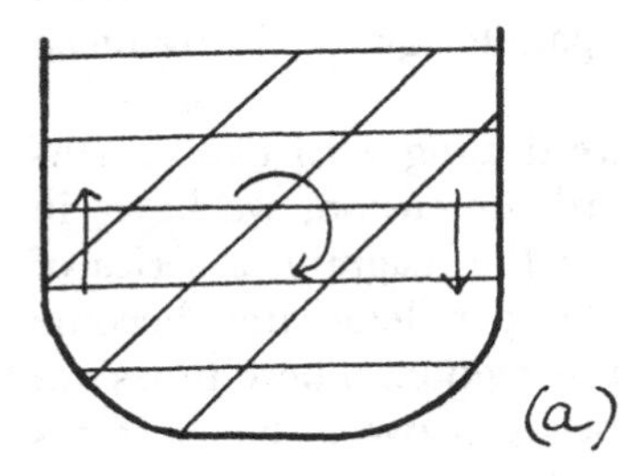

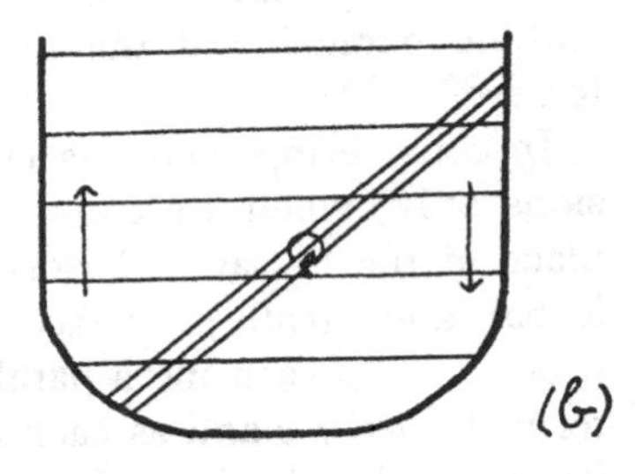

Fig. 38.

of the centrifugal force owing to the earth's rotation: this varies with the sine of the latitude; (ii) the difference in density between the oil and the water; and (iii) the velocity with which the oil is moving along the tank. If the substratum of water were also moving, this factor affecting the obliquity of the interface would be the difference in velocity between the oil and water layers. Given a steady motion, these factors adjust themselves in a perfectly definite way, with the interface usually lying at an angle to the horizontal.

Turning from this illustration to the conditions in an actual mass of water, where the surfaces of equal density are sloping and where there is no force, such as that set up by the rotation of the earth, keeping the denser water heaped up to one side, the water will shift in its effort to regain equilibrium and the sloping surfaces become horizontal. Fig. 38 (*a*) shows, in section, sloping surfaces of equal specific volume in such a mass of water. The specific volume, v, is the volume occupied by one gram of the water, and is equal to $1/\rho$ where ρ is the density.[1] The sections of the surfaces of equal specific volume are known as *isosteres*. The horizontal lines in the diagram depict the sections of the surfaces of equal pressure or *isobars*, which for the present purpose approximate sufficiently to the lines of equal depth. The arrows show the rotary movement of the water which would be occasioned by the forces due to the distribution of the density or related specific volume.

The measure of these forces is represented by the number of parallelograms formed by the intersection of the isosteres and isobars in a hydrographic section. After the movement has taken place and equilibrium been established, the isosteres and isobars will lie parallel and the number of parallelograms be reduced to zero.

In considering a cross-section we are dealing with two dimensions only; when we consider the third dimension, we have, in place of the isobar and isostere lines in the diagram, a series of isobar and isostere surfaces, while the parallelograms become tubes, each presenting a parallelogram in section. These tubes are termed *solenoids* and as each one lies between two isobar surfaces its course lies horizontal. Since the isobar surfaces and isostere surfaces cannot terminate in the water itself but must continue until reaching a bounding surface, so the solenoid tubes must also continue until they reach either a boundary surface of the water-mass, or else turn back upon themselves.

In order to measure the forces giving a rotary movement to the water in C.G.S. units it would be convenient if possible to draw the isosteres one C.G.S. unit apart, but the differences in specific volume in the sea are only a fraction of unit specific volume. If the isosteres are drawn as in Fig. 39 for every 10^{-5} unit of specific volume and the isobars for every 10^{5} unit of pressure, then each

[1] When dealing with considerable depths, the effect of pressure upon the specific volume of the water should be allowed for. Given the salinity and the temperature *in situ*, the density at that temperature can be obtained from Knudsen's *Hydrographical Tables*. Tables to convert the specific volume, or reciprocal of the density, at atmospheric pressure into the specific volume *in situ* at different depths have been published by Hesselburg and Sverdrup, by Smith and by Sund.

parallelogram represents one unit C.G.S. solenoid. It so happens that a pressure of very nearly 10^5 C.G.S. units is produced by a head of 1 metre of water. In actual practice it may be more convenient to draw the isobars still further apart; if 10 metres apart, then each parallelogram represents the section of a 10 C.G.S. unit solenoid.

In the attempt of the isostere surfaces to become horizontal, each such solenoid will produce the same rotary movement in the water and the number of solenoids is therefore a measure of the intensity attained by the forces causing the water to rotate.

When a vessel of water as depicted in section in Fig. 39 attains equilibrium, the water ceases to rotate, and, the isosteric surfaces lying parallel with the isobars, the number of solenoids becomes zero.

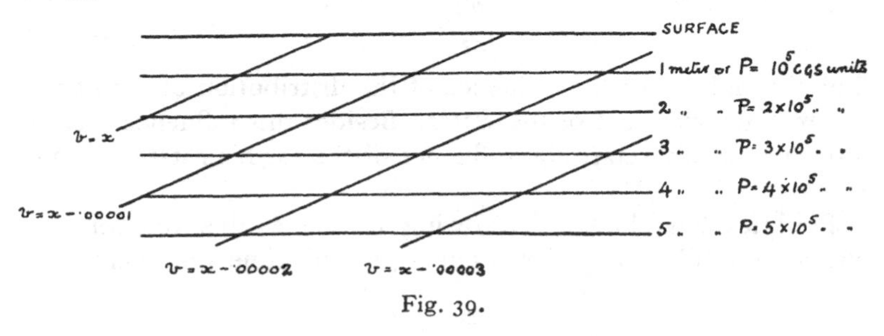

Fig. 39.

In order to calculate the effect of the solenoids on the rotary movement of the water, Bjerknes studies the movement of a series of water-particles which compose a ring or closed curve. He calculates the tangential velocity of these particles, following the whole curve once round. The sum of these tangential velocities is termed the *Circulation of the Curve*—*C*.

In order to envisage the term *C*, consider a number of water-particles which form a circle, the circle being vertical and at right angles to a current of water which is moving faster at the surface than at the bottom. After a short period of time the circle will be drawn out to an oval—no longer vertical to the surface. The particles on the 'sides' of the oval which are continuously lengthening are consequently increasing in their tangential velocity, and the value of *C* for the curve is continually increasing.

The translatory movement of the water, or current, is not

discernible in the 'circulation of the curve' (C), the value of which is a measure of the rotary movement only.

Thus a closed curve in a current, either slow or rapid, where all the particles of water have the same forward velocity, will have C = zero, since there are an equal number of particles with tangential velocities in opposite directions.

It is only when the translatory movement of the water is of such a nature that it distorts a curve of water-particles within it that C will have a value other than zero.

Then the value of C is not constant, but is continually augmented by the rotary tendency of the solenoids. According to Bjerknes, the increase in C per second is equal to the number of solenoids, A, within the closed curve or

$$\frac{dC}{dt} = A \qquad \text{......(1).}$$

This formula gives the influence of the distribution of density on the rotary movement of the water. Besides this influence there is also the, usually retarding, influence of the earth's rotation and of friction.

Bjerknes has shown that owing to the earth's rotation, the alteration in C for any closed curve per unit time amounts to

$$2\omega \frac{dS}{dt},$$

where S is the area of the closed curve when projected upon the equatorial plane and ω is the angular velocity of the earth's rotation.

If R represents the influence of friction upon the movement, then

$$\frac{dC}{dt} = A - 2\omega \frac{dS}{dt} - R \qquad \text{......(2).}$$

This formula contains all that influences the rotary circulation of the water in the sea.

From this formula Sandström has calculated the velocity of *translatory* movements in currents in the following manner. In equation (2) the two terms $\frac{dC}{dt}$ and R are small in comparison with the intermediate ones, and for a first approximation are disregarded, when

$$A = 2\omega \frac{dS}{dt} \qquad \text{......(3).}$$

Since it is desired to measure horizontal velocities and not the comparatively insignificant vertical currents, the change in area of the projected curve per second $\left(\frac{dS}{dt}\right)$ due to the horizontal current is sufficient. This then becomes $\frac{ds}{dt} \times$ sine latitude, where s is the area of the curve projected on the surface of the sea, and $\frac{ds}{dt}$ is the change in its area per second.

Equation (3) now becomes

$$A = 2\omega \frac{ds}{dt} \text{ sine latitude} \qquad \ldots\ldots(4).$$

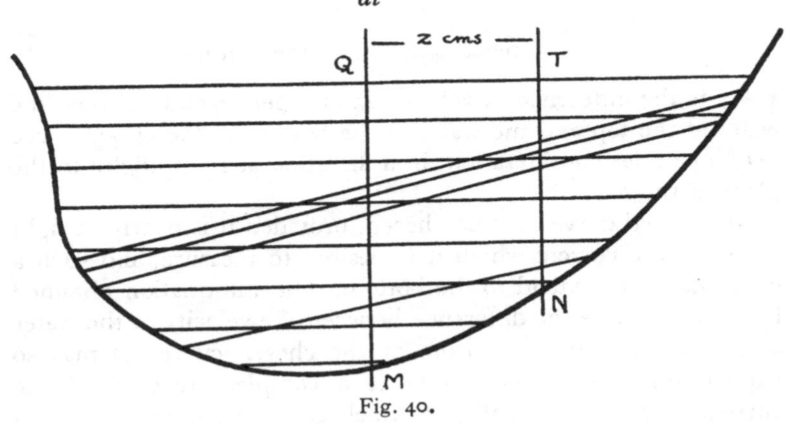

Fig. 40.

A hydrographical section (vertical to the sea's surface) is drawn as nearly as possible *at right angles to the current* (Fig. 40). Isosteres and isobars are drawn at convenient intervals (in powers of 10 C.G.S. units) and the number of solenoids counted in a section between two hydrographical stations, several, or even many kilometres (z cm.) apart.[1]

[1] The number of solenoids, A, in a rectangular section between equal vertical depths at two positions at sea may also be calculated:

A = depth in metres $\times 10^5 (v_{m_1} - v_{m_2})$ C.G.S. solenoids,

where

v_{m_1} = the mean specific volume of the water between the upper and lower depths at the one position;

v_{m_2} = the mean specific volume of the water between the same upper and lower depths at the other position.

If we now consider a closed curve falling between two stations such as *QTNM* containing A C.G.S. solenoids, and where the surface water is moving at x cm. per second in a direction at right angles to the plane of the paper, while the bottom water is moving only at y cm. per second, then at the start the area of the closed curve projected on the surface of the sea is zero, but after 1 second it is $z(x-y)$ sq. cm. That is $\frac{ds}{dt} = z(x-y)$.

From equation (4)

$$x - y = \frac{A}{2\omega \times z \times \text{sine latitude}},$$

or

$$x - y = \frac{A}{0{\cdot}0001458 \times z \times \text{sine latitude}} \qquad \ldots\ldots(5),$$

$x - y$ is the difference in velocity in cm. per second between the water at the top and the water at the bottom of the closed curve *QTNM* under consideration, in a direction at right angles to the plane of the paper.

Any closed curve may be chosen, provided it is nearly at right angles to the current which it is desired to measure, and such a curve need not extend to the bottom. The information obtained by applying (5) is the difference between the velocity of the water at the top and at the bottom of the chosen curve. It may so happen that the isosteres indicate a complete reversal of the current at a certain depth, in which case a number of 'closed curves,' one above the other, can be chosen to give values showing the change in velocity and direction with increasing depth.

As a simple case for the application of equation (5) consider an ocean current running over a substratum of motionless water in the northern hemisphere. Owing to the rotation of the earth, the current will veer to the right until it encounters a coast, which it will then follow, still pressing towards the right, that is, setting in towards the coast. In consequence of this pressure the current will become deeper near the coast than farther out; the separating surface between the current and its substratum will lie obliquely. In this surface, a number of solenoids will lie in the longitudinal direction of the current. By taking a hydrographical section at right angles to the current and applying equation (5), the speed of the current in cm. per second is arrived at, since it is running over still water. If the section is not taken at right angles to the current, the component of the current's velocity at right angles to the section is obtained. The direction of the current is indicated by the lighter surface water being pressed over to the right when facing in the direction of the current in the northern hemisphere.

This estimation is subject to a source of error which may occur at times. An on-shore wind, at right angles to the current, will tend to pile up the lighter water against the coast, when the velocity calculated from the solenoids will be too high. With an off-shore wind the reverse is the case.

In order to make the practical use of this theory clear, the following illustration is taken. In the course of the annual survey of the currents south of the Grand Banks by the Atlantic Ice Patrol, temperatures and salinity of the water at various depths down to 750 metres were observed in 1922 at two positions 59 kilometres apart on a line at right angles to the east-going Gulf Stream current. It was desired to calculate the velocity of the current at the surface, and since it is known that the greatest variations in density of the sea occur in the upper levels, it was *assumed* that at 750 metres the velocity at which the water moved was negligibly small.

At the northern station the following observations of temperature were made at the different levels, the salinities of the samples of water collected were determined, and from the temperature *in situ* and the salinity the specific volume was calculated, allowance being made for the effect of pressure.

Table XXXVII. Northern Station (near northern limit of Gulf Stream).

Depth in metres	Temp. °C.	Salinity parts per 1000	Specific volume corrected for effect of pressure
0	5·7	33·93	0·97393
50	12·0	35·31	0·97363
125	10·1	35·16	0·97312
250	6·7	35·00	0·97217
450	5·4	35·04	0·97119
750	4·6	35·01	0·96975

From this the mean (integral) specific volume of the water above 750 metres was found to be 0·97163.

Table XXXVIII. Southern Station (in east-going Gulf Stream).

Depth in metres	Temp. °C.	Salinity parts per 1000	Specific volume corrected for effect of pressure
0	18·1	36·21	0·97449
50	18·0	36·33	0·97414
125	16·3	36·11	0·97363
250	12·9	35·56	0·97278
450	9·2	35·12	0·97159
750	6·6	34·95	0·97005

The mean (integral) specific volume of the water down to a depth of 750 metres at this station was found from the above values to be 0·97211.

In the section 750 metres deep between the two stations, the number of C.G.S. solenoids, A, is therefore equal to $750 \times 10^5 (0{\cdot}97211-0{\cdot}97163)$ or 36,000 C.G.S. solenoids (see footnote, p. 127). From formula (5) velocity in cm. per second $= x - y = \dfrac{A}{0{\cdot}0001458 \times z \times \text{sine latitude}}$. In this case z, the distance between the stations $= 59 \times 10^5$ cm., and the sine of the mean latitude 41° 10′ N. = 0·658. Hence velocity

$$= \frac{36{,}000}{0{\cdot}0001458 \times 59 \times 10^5 \times 0{\cdot}658} = 63\tfrac{1}{2} \text{ cm. per sec.}$$

or 1·3 nautical miles per hour.

BIBLIOGRAPHY

TIDES AND TIDAL STREAMS

WARBURG, H. D. *Tides and Tidal Streams*. Cambridge, 1922.

MARMER, H. A. *The Tide*. New York, 1926.

AYLMER, E. F. and WHITE, J. *Admiralty Manual of Navigation*. London, 1915.

TIZARD, T. H. *Tides and Tidal Streams of the British Isles*. London, 1909.

WERENSKIOLD, W. "An Analysis of Current Measurements in the Open Sea." *Fordhandlinger ved de Skandinaviske naturforskeres* **16***de mote i Kristiania* 10. Oslo, 1918.

OCEAN CURRENTS

(1) NELSON, E. W. "The Manufacture of Drift Bottles." *Journ. Mar. Biol. Assoc.* **12**, 700–16. 1922.

(2) CARRUTHERS, J. N. "The Water Movements in the Southern North Sea. Part II. The Bottom Currents." *Fishery Investigations*, Series 2, **9**, No. 3. London, 1926.

(3 *a*) HELLAND-HANSEN, B. "The Ocean Waters." *Internat. Rev. d. Hydrobiol. u. Hydrog.* 1911 and 1923.

(3 *b*) KRUMMEL, O. *Handbuch der Ozeanographie.* Stuttgart, 1911.

(3 *c*) MATTHEWS, D. "Physical Oceanography." *Dict. Applied Physics*, **3**. 1923.

(4) EKMAN, V. "Earth Rotation and Ocean Currents." *Archiv f. Matematik, Astron. och Fysik*, **2**, No. 11. Uppsala, 1905.

"Über horizontale Zirculation bei winderzeugten Meereströmungen." *Ibid.* **18**, 26. 1923.

LAMB, H. *Hydrodynamics*, p. 561. Cambridge, 1924.

JEFFREYS, H. *Phil. Mag.* **39**, 578–85. 1920.

SVERDRUP, H. *Journ. Washington Acad. Sci.* **16**, 529–40. 1926.

(5) GARSTANG, W. "Report on the Surface Drift in the English Channel during 1897." *Journ. Mar. Biol. Assoc.* **5**, 199. 1898.

CARRUTHERS, J. N. "Investigations upon the Water Movements in the English Channel." *Journ. Mar. Biol. Assoc.* **14**, 685–722. 1927.

(6) —— "Water Movements in the Southern North Sea. Part I. The Surface Drift." *Fishery Investigations*, Series 2, **8**, No. 2. 1925.

(7) SCHOTT, G. *Physische Meereskunde*, p. 64. Berlin, 1924.

(8) WITTING, R. "Kenntnis des vom Winde erzeugten Obenflächenstomes." *Ann. Hydrog.* p. 193. 1909.

—— "Die Meeresoberfläche, die Geoidflache und die Landhebung dem Baltischen Meere entlang und an der Nordsee." *Fennia*, **39**, No. 5. Helsingfors, 1918.

—— "L'Influence de l'État de l'Atmosphère sur la Surface de la Mer." *Oversigt af Finska Vetenskaps-Soc. Förhand.* **59**, 1916–1917, Afd. A, No. 13. Helsingfors, 1917.

(9) DOODSON, A. T. "Meteorological Perturbations of Sea Level." *Monthly Notices Roy. Astr. Soc. Geophysical Suppl.* I. 124. 1924. Also *Nature*, **112**, 765. 1923.

CRESSWELL, M. "Tides and Currents, and the Effect of the Wind on the Water Level near the Shore with Set and Drift associated." *Marine Observer*, **3**, 137. August 1926.

(10) EKMAN, V. *Publ. de Circonstance*, No. 43. 1908.

—— *Ibid.* No. 49. 1910.

(11) HESSELBURG and SVERDRUP. "Die Stabilitätsverhältnisse des Seewassers." *Bergens Museums Aarbok*, 1914–1915. No. 15.

(12) SANDSTRÖM, J. W. *Archiv f. Matematik, Ast. och Fysik*, **9**, No. 32, p. 2. 1913.

(13) PETTERSSON, O. "On the Influence of Ice melting upon Oceanic Circulation." *Svenska Hydrog. Biol. Komm. Skrifter*, **2**. 1904. *Geograph. Journ.* **24**, 285, 1904, and **30**, 273, 1907.

SANDSTRÖM, J. W. "On Ice melting in Sea Water and Currents raised by it." *Svenska Hydrog. Biol. Komm. Skrifter*, **2**. 1905.

AITKEN, J. "The Melting of Icebergs." *Nature*, **90**, 513. 1913.

(14) NANSEN, F. "Waters of the N. Eastern N. Atlantic." Leipzig, 1913

(15) Harvey, H. W. "Water Movement and Sea Temperature in the English Channel." *Journ. Mar. Biol. Assoc.* **13**, 659–64. 1925.

(16) Helland-Hansen, B. "Physical Oceanography" in Murray and Hjort, *Depths of the Ocean*, p. 304. London, 1912.

(17) Bowman, A. "Biological Exchanges between the Atlantic and N. Sea." *Fish. Scot. Sci. Invest.* 1922, No. 11. Edinburgh, 1923. Also *British Assoc. Report*, p. 367. 1922.

Schmidt, J. *Rapp. et Proc. Verb. du Cons. Internat.* **10**, 159. 1909.

Hardy, A. C. *Publ. de Circ.* No. 78. 1923.

(18) Helland-Hansen, B. and Nansen, F. "The Eastern North Atlantic." *Geofysiske Publikasjoner*, **4**, No. 2. Oslo, 1926.

(19) Merz, A. "Die Deutsche Atlantische Expedition auf dem Vermessungs- und Forschungsschiff 'Meteor'." *Sitzber. der Preuss. Akad. der Wissenschaften*, **31**, 562–86. 1925.

(20) Sandström, J. W. "The hydrodynamics of Canadian Atlantic Waters." *Canadian Fish. Exped.* 1914–1915. Ottawa, 1918.

(21) Pettersson, O. "Hydrography in the Transition Area." *Journ. du Cons. Internat.* **1**, 305–321. 1926.

BJERKNES' CIRCULATION THEORY AND ITS APPLICATION

Bjerknes and Sandström. *Dynamic Meteorology and Hydrography*. Carnegie Institute.

Sandström, J. W. *The Hydrodynamics of Canadian Atlantic Waters*. Canadian Fish. Exped. 1914–1915. Ottawa, 1918.

Sandström and Helland-Hansen. "On the Mathematical Investigation of Ocean Currents." Translated by O. W. Thompson. *North Sea Invest. Committee's Report on Fish. and Hydrog. Investigations*, 1902–1903 (C.D. 2612), publ. 1905.

Hesselburg and Sverdrup. "Beitrag zur Berechnung der Druck und Masseverteilung im Meere." *Bergens Museums Aarbok*, 1914–1915, No. 4.

Smith, E. H. "A Practical Method of Measuring Ocean Currents." *U.S. Coast Guard Bulletin*, No. 14. Washington, 1926.

Sund, O. "Graphical Calculation of Specific Volume and Dynamic Depth." *Journ. du Cons. Internat.* **1**, 235–41. 1926.

Chapter IV

TEMPERATURE OF THE SEA

The considerable effect of relatively small changes of temperature on the biological processes of marine plants and animals has been discussed, and it is apparent that the temperature of the water is a factor of primary importance in its effect on the population. For this reason, the study of the conditions controlling the distribution of temperature in the sea, its annual variation or range, and the fluctuations which occur from year to year, is a facet of physical oceanography of equal importance to the biologist as that of horizontal and vertical currents.

GAIN AND LOSS OF HEAT

There is a gain in heat from the energy of solar radiation which has penetrated the atmosphere and fallen upon the surface of the sea. This radiation is a mixture of the rays of sunlight coming direct from the sun and of rays reflected from the sky. As oblique rays enter the water they are refracted and pass through the water at an angle to the horizontal greater than in their passage through the atmosphere, and at an angle never less than the critical angle. A certain amount of the solar radiation is reflected from the surface of the sea, the amount increasing as the path of the rays becomes more nearly horizontal. Birge has calculated that if equal quantities of solar radiation came from all areas of the sky, the loss due to reflection would amount to about 17%. As the major portion of the daily solar radiation occurs in the middle of the day, when the radiation comes more from overhead, the mean loss is presumably less than this value except in high latitudes.

The incident solar radiation is made up of rays of varying wave length. The greater portion have lengths between 0·4 and 0·7 μ*. As these pass through water the rays of longer wave length are most rapidly absorbed. On travelling through a metre of pure water about half of the energy of the rays which enter the water is

* μ = ·001 millimetre.

absorbed; on travelling through a further metre of pure water about 20 % more of the energy is absorbed, not 25 %, since the rays which have already penetrated the first metre contain a greater proportion of shorter wave length rays.

In the sea and in lakes rays are absorbed and reflected by suspended particles and plankton organisms in addition to being absorbed by the water itself. The absorption of energy with depth is more rapid than in pure water.

We have a considerable knowledge of the rate of absorption of rays of about 0·45 μ on their passage through sea water. There are no direct measurements, so far as the writer is aware, of the penetration of *energy*, due to rays of all wave lengths, into the sea, but in several American lakes measurements have been carried out by Birge and his co-workers (7). The results of these measurements in the clear water of Lake Seneca are very striking. Approximately 80 % of the energy which enters the water is absorbed in the upper metre layer. From the observations it would appear that the major part of this 80 % loss occurs in the upper 20 cm. About 5 % reaches a depth of 5 metres and 1 % a depth of 10 metres. Birge remarks that a considerable increase in transparency of the water would leave these figures, not unchanged, but of the same order of magnitude. Since the percentages arriving at various depths are based on observations made in direct sunlight with the sun high in the sky, they are maximal values, because with diffuse light from the sky rays would on the average have to travel further before reaching the 1, 5 and 10 metre levels.

Motionless water is a very poor conductor of heat, but, in the sea, the ceaseless movement of water-particles up and down past one another causes heat to be transferred downwards in precisely the same way as the momentum of horizontally moving particles is carried downwards.

The transference of heat downwards to lower levels is controlled by the amount of turbulence or eddy motion, unless there is an actual redistribution of the water itself, brought about by currents.

The rate of change of temperature at a distance y below the surface is shown in the following equation

$$\frac{\delta\theta}{\delta t} = c\,\frac{\delta^2\theta}{\delta y^2},$$

where θ is the temperature, t time and c is the virtual coefficient of thermal conductivity which is numerically equal to the virtual coefficient of viscosity (p. 102).

Where warm water rests on colder water below, eddy motion is restrained at the junction owing to differences in density, and heat can only pass the junction slowly.

The amount of solar radiation falling on a horizontal square centimetre at various latitudes is given in the following table, the amount per day falling on the equator when the sun is directly overhead at midday being taken as unity.

Table XXXIX.

		0°	20° N.	40° N.	60° N.	90° N.
Vernal equinox	March 20	1·000	0·934	0·763	0·499	0·000
Summer solstice or midsummer in N. latitudes	June 21	0·881	1·040	1·103	1·090	1·202
Autumn equinox	Sept. 22	0·984	0·934	0·760	0·499	0·000
Winter solstice	Dec. 21	0·942	0·679	0·352	0·000	0·000
Annual total ...	—	347	329	274	197	143

This table shows that there is more solar radiation received during the long days in the height of the summer in northern latitudes than during the '12-hour day' in the tropics.

The number of hours of sunshine has a preponderating influence upon the whole day's solar radiation. From observations carried out by Ångström ((1) 1924), the total solar radiation found at Stockholm and Washington approximates very nearly to

$$Q_s = Q_o\,(0{\cdot}25 + 0{\cdot}75\,S),$$

where S = the proportion of the actual to the possible hours of sunshine during the day,

Q_o = solar radiation corresponding to a perfectly clear day,

Q_s = solar radiation recorded at the place of observation (mean value for each month).

The amount of solar radiation falling on the sea at any moment varies with the angle of incidence of the sun's rays and to a very marked extent with the atmospheric conditions.

In addition to receiving heat from solar radiation a particular area of the sea may gain heat by the inflow of warmer water. Water masses differing 10° C. in temperature are on occasions separated by only a few yards. The warm Atlantic water which passes north through the wide and deep gap between Scotland and Iceland brings with it an almost temperate climate on the south coast of Iceland at sea level. The temperature during some winters never falls below zero Fahrenheit and fifteen below is more often experienced in New York than at Reykjavik, its capital town. The Atlantic water creeps north and melts away the ice which would otherwise be there, so that whalers can sail from six to seven hundred miles closer to the Pole on the Atlantic side of North America than on the Pacific side, where the Japan current cannot pass in any quantity into the Arctic Ocean through the shallow Bering Strait which is only 40 miles wide.

The sea as a whole loses heat by evaporation and by radiation outwards. The amount lost by directly heating the atmosphere is relatively small in quantity, since the heat lost by one cubic metre of water in falling 1° C. will heat a column of air one square metre in cross-section and 3000 metres high through 1° C. Owing to its high specific heat the sea acts as a very efficient temperature regulator to the atmosphere and adjacent land.

The radiation outwards from the sea into space is of wave length about 6 to 10 μ, over ten times the wave length of the incoming radiation, and for such the atmosphere acts as a blanket. If it were not for the atmosphere stopping emission, the loss by radiation would greatly exceed the gain by solar radiation until equilibrium were attained. Under this hypothetical condition water at 17° C. would emit 833 grm. calories per day per sq. cm.,* whereas the solar radiation penetrating the atmosphere and falling on 1 sq. cm. in London during the summer rarely exceeds 400 grm. calories per day.

With a clear sky 25 % (to a possible maximum of 50 %, with a dry atmosphere) of σT^4 may be emitted and lost into space, but with an overcast sky it is not more than a tenth of this amount (1).

* The rate of emission of energy into space by a 'black' body at $T°$ Absolute and not surrounded by an atmosphere is taken as σT^4 ergs per second, where $\sigma = 5{\cdot}72 \times 10^{-5}$.

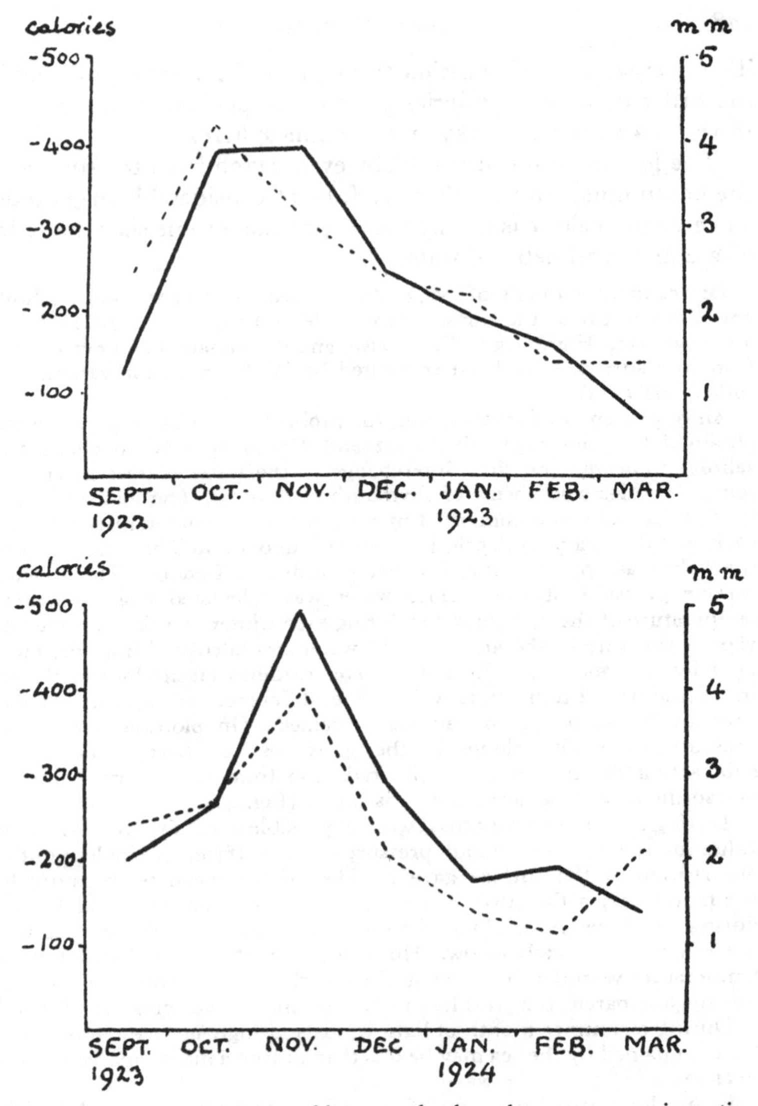

Fig. 41. Thick line shows loss of heat per day by column 1 sq. cm. in section, from surface to bottom, at lat. 50° 02′ N., long. 4° 22′ W., *minus* solar radiation per horizontal sq. cm. as recorded at South Kensington. Dotted line shows difference in aqueous vapour pressure between surface water at same position and the air at Cattewater Air Station Observatory.

The average rate of emission throughout the year is probably in the order of 80 grm. calories per sq. cm. per day, more perhaps in regions with a clear sky and dry atmosphere.

The loss of heat occasioned by evaporation from the surface of the ocean must, on the other hand, be of considerable magnitude, since 1 grm. calorie is lost by the evaporation of one six-hundredth of a cubic centimetre of water.

Direct measurements of evaporation of sea water of the same salinity and temperature as the surface were made by Lütgens ((2a) 1911) during a voyage from Hamburg to Valparaiso, and the radiation and evaporation from the surface have been computed by W. Schmidt for various latitudes ((2b) 1915).

An opportunity of investigating the probable effect of evaporation was obtained between September 1922 and March 1924 when changes of salinity indicated but little interchange of the water masses in an area lying 20 miles south-west of Plymouth ((2c) 1925). The loss or gain in heat of the water was computed by integrating the temperatures recorded each month at varying depths from surface to bottom. The solar radiation was taken as approximating to that recorded in London. The aqueous vapour pressure of the surface water was calculated from the mean temperature of the surface water during each winter month. The aqueous vapour pressure of the air above the water was calculated from the mean of daily wet and dry bulb thermometer readings taken close to the sea in Plymouth. From these values the difference in aqueous vapour pressure between sea and air was obtained. On plotting this vapour pressure difference alongside the *gross* loss of heat, obtained by subtracting the gain due to solar radiation from the observed loss, an extraordinarily close agreement was found (Fig. 41).

During the summer months it was not possible to obtain a representative value for the aqueous vapour pressure at the surface. A single monthly observation at the surface gave no idea of the mean temperature for the month, since the surface was subject to more or less rapid heating during sunshine and cooling when wave motion mixed the surface with the layers immediately below. However, on plotting the distribution of temperature with depth, it was noticeable that when a warm upper layer was most apparent the *gross* loss in heat by the sea was greatest (Fig. 42).

During a summer month of light winds and high surface temperature, the heat gained by the sea may be less than during a month marked by less solar radiation and more wave motion.

Before leaving the question of exchange of heat, mention should be made of the mathematical treatment of this subject by McEwen, who computed the mean monthly surface temperature at various latitudes in the Pacific and found them to be in very fair agreement with the observed values ((2d) 1918). By inserting a term expressing the rate of change of heat due to horizontal currents into the equation, and applying the formula to two regions where horizontal currents exist, estimates of the horizontal flow were arrived at in good agreement with direct observations. This research is still in progress at the Scripp's Institute, and further reports are imminent.

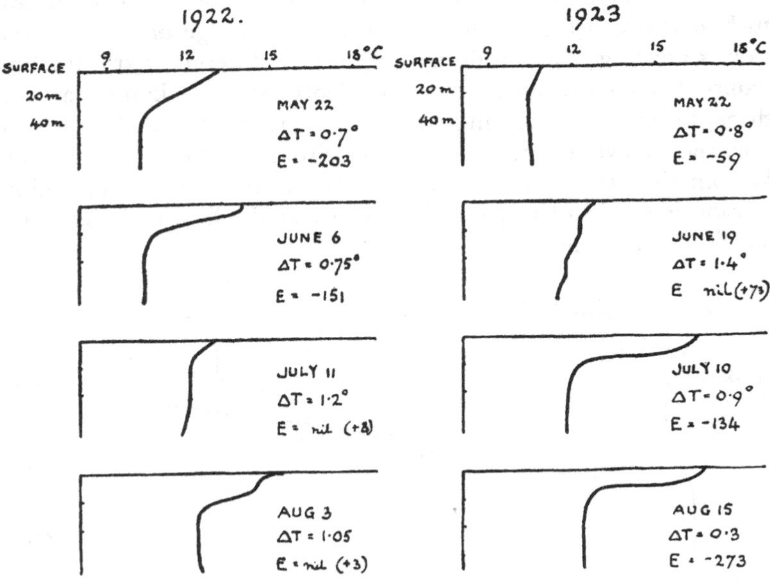

Fig. 42. *E*, the gross daily loss of heat, being the loss of heat per day by a column 1 sq. cm. in section, *minus* solar radiation per horizontal sq. cm. as recorded at South Kensington. Black line shows the distribution of temperature with depth at lat. 50° 02′ N., long. 4° 22′ W. during the summers 1922, 1923. ΔT shows the change in temperature of the whole column of water, as if mixed, during the calendar month.

DISTRIBUTION WITH DEPTH

Decreasing temperature with increasing depth is usual in the deep open oceans, except during winter in high latitudes. The curves *A*, *C* and *D* in Fig. 43 show typical instances.

As the coasts are approached in temperate latitudes and as the depth falls to 100 metres or less, the cooling of the surface water during the autumn months gives rise to convection currents. The colder and heavier surface water sinks and brings about vertical mixing. The breakdown of the layering allows eddy motion due to waves to extend deeper, and this helps, until finally almost the same temperature is attained from top to bottom; this occurs in the English Channel in October or November. In such localities in summer the surface layers become heated to a greater extent than the deeper water. At a certain depth there is insufficient turbulence to bring about the downward passage of heat as fast as it is arriving. The resulting temperature differences still further damp down eddy motion at this depth—a zone known as the discontinuity layer, sprungschicht or thermocline—and heat can only pass slowly to the water below. From the thermocline to the bottom the water remains more or less isothermal because eddy motion is sufficient to pass on downwards the heat arriving quite slowly through the thermocline.

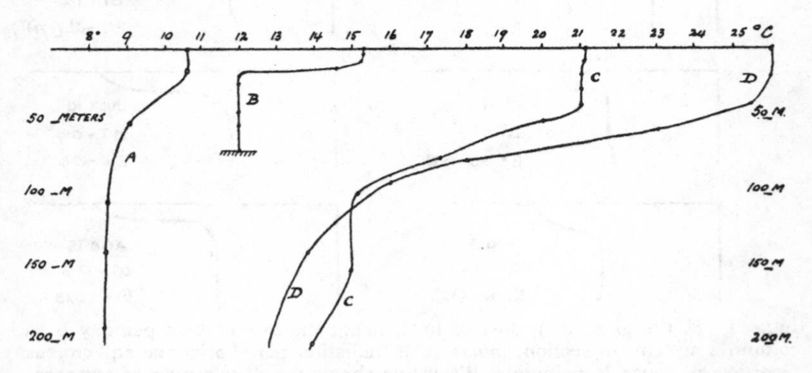

Fig. 43. Distribution of temperature with depth. *A*. Lat. 57° 17′ N., long. 21° 43′ W., depth over 900 m., *Fridjhof*, July 9, 1910. *B*. Lat. 50° 02′ N., long. 4° 22′ W., in English Channel, depth 71 m., *Salpa*, Aug. 7, 1924. *C*. Lat. 37° 44′ N., long. 13° 21′ W., depth over 2000 m., *Discovery*, Oct. 12, 1925. *D*. Lat. 7° 01′ N., long. 15° 55′ W., depth 2425 fathoms, *Challenger*, Aug. 16, 1873.

The depth of the upper stratum of warm water is not very great, and at times it is very sharply divided from the colder water below by the discontinuity layer (Fig. 43, *B*). The two strata have been termed by Atkins the *Epithalassa* and *Hypothalassa*. Under these conditions the sea may be considered, from the point of view of the cold-blooded animals and phytoplankton living in it, as divided into two absolutely distinct layers. In the upper warm stratum their rate of metabolism or of respiration may be over 50 % more rapid than in the water below.

Table XL. Distribution of temperature with depth. English Channel. July 19, 1923. Lat. 50° 02′ N., long. 4° 22′ W.

Depth (metres)	Temp. °C.	
0	16·15	Epithalassa
5	16·08	
10	15·85	
12½	15·82	
15	15·82	
		Discontinuity layer or thermocline
17½	12·09	Hypothalassa
20	12·05	
30	12·05	
60	12·03	

Temperature layering commences about May and lasts until sometime in September. Several days of strong winds or a gale affects the layering by wave action, causing the warmer water to extend to a greater depth and the temperature gradient to be less steep or sudden (Fig. 44).

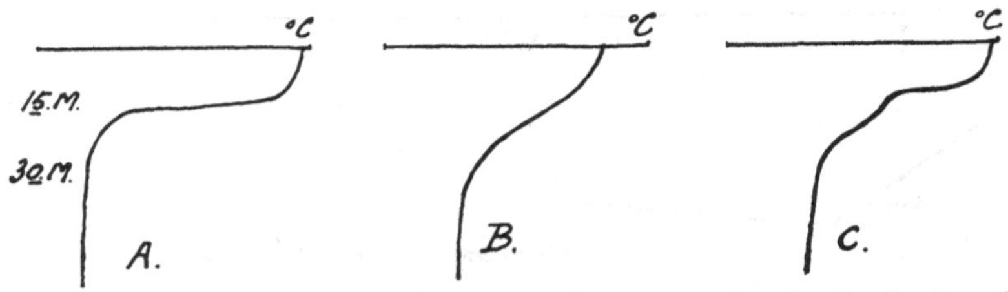

Fig. 44. Diagram showing the effect of a summer gale on the distribution of temperature with depth, from observations in the English Channel. *A*. After hot calm weather. *B*. After a gale. *C*. Later after a resumption of calm sunny weather.

As the temperature of sea water increases, the difference in density due to a further unit rise of temperature becomes very much greater.

On rising from 0° C. to 2° C. the decrease in density is 0·00013.
" 10° C. to 12° C. " 0·00036.
" 20° C. to 22° C. " 0·00054.

Hence a temperature gradient of, for instance, 0·0001° C. per cm. will be more stable and require more work to break it down in a warm sea of homogeneous salinity than in a cold sea ((3) 1925).

It is only when the surface temperature in autumn falls to near the temperature below the discontinuity layer that the layering will be broken down and mixing of the hypothalassa with the epithalassa by convection currents take place to any extent. In consequence, the deeper strata reach a maximum temperature each year after the upper warm layers have commenced to cool. In the English Channel the surface temperature well off-shore is at a maximum about August, while the deep water ordinarily

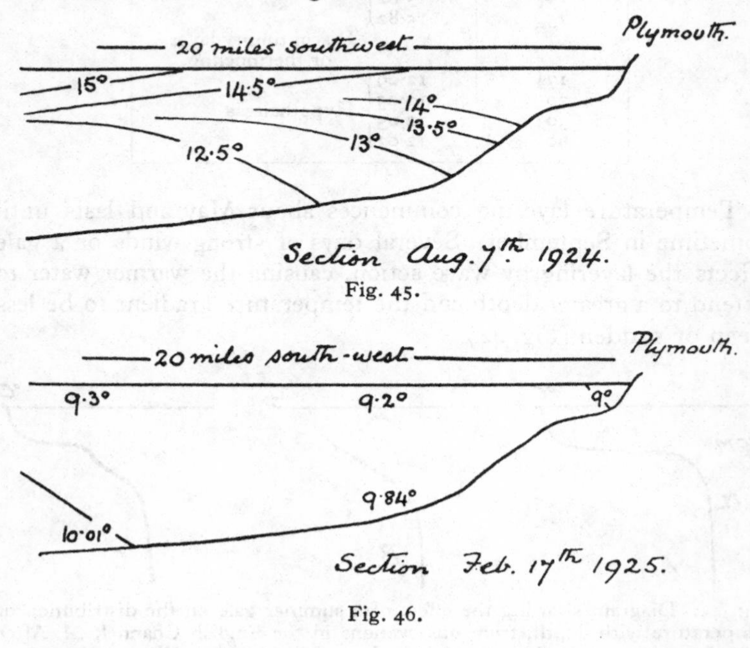

Fig. 45.

Fig. 46.

attains its greatest temperature in October, by which time the convection currents can penetrate to the bottom. In areas of greater depth, the deep water may not attain its maximum temperature until the following year. In fact a reversal of the seasons occurs in deep water layers in such cases.

As the coast is approached from seaward with gradually decreasing depth, and inshore conditions are reached, the tidal currents tend to break down the distinct layering which occurs further out to sea. This is shown very clearly in Fig. 45. The

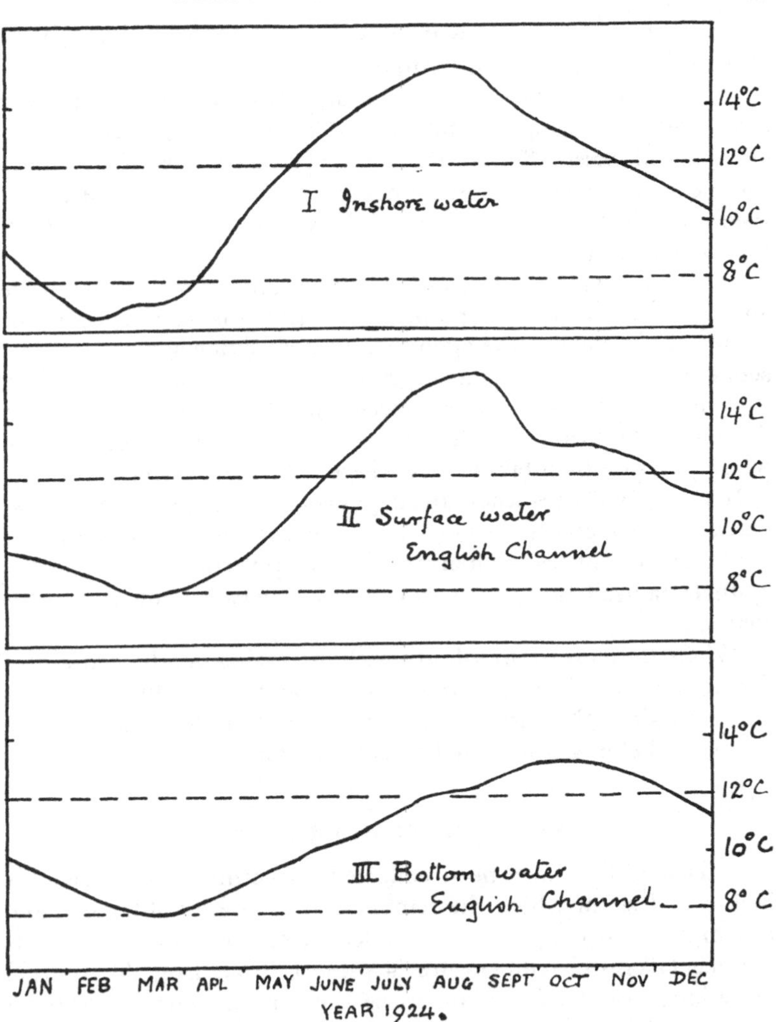

Fig. 47. Annual variation in temperature of inshore water, at a position in Plymouth Sound, of surface and of bottom water in the English Channel, lat. 50° 02′ N., long. 4° 22′ W., depth 70 metres.

inshore water is much warmer than the deeper water to seaward, but the surface does not attain such a high temperature, except in very shallow pools, etc., sheltered from the tidal streaming.

Turbulence and mixing will be accentuated where strong tidal streams run over a rough bottom.

In addition to tidal streams, the piling up of water against the coast and its flowing out to seaward as a bottom or subsurface current will cause effective mixing. Off-shore winds which lower the mean sea level by blowing a surface current out to sea, have the same effect, since the place of the water running out is taken by water from below.

In the relatively shallow *inshore* waters on our temperate coasts animals have no cool hypothalassa during the summer months such as they have in the open sea. Bottom-living species are consequently exposed to a much greater *range of temperature* during the year, and this is further accentuated because loss of heat during winter from the shallow water gives rise to a greater fall in temperature than takes place where the water is deeper.

In the open oceans where the depth is great, the warm epithalassa extends considerably deeper. Here wave motion extends further down and doubtless submarine waves help vertical mixing in the epithalassa itself, and to some small extent between epi- and hypothalassa.

Where an increase in salinity with increasing depth occurs, this further restricts the transference of heat and mixing by convection currents, since the surface water will have to become colder than the water below before equal density is attained.

SURFACE TEMPERATURE

Extensive data have been collected concerning the surface six inches or less, of which the temperature is readily obtained from a sample dipped up with a wooden or leather bucket. In this surface layer the temperature is subject to an irregular diurnal variation; in hot calm summer weather around our coasts there may be a rise of 2° C. between dawn and the early afternoon, and after a frosty night in November the surface has been noticed to be 0·7° C. colder than at a depth of 5 metres. In fresh-water ponds during sunny weather the temperature gradient in the top few inches is very steep; in the sea wave motion breaks this down to a large extent, but not entirely. After a 'heat wave' and in calm weather some two miles off the Cornwall coast the writer has observed a difference of 1¼° C. between the water about an inch below the surface and at a depth of 10 to 12 inches.

In areas such as the southern part of the North Sea and the eastern end of the English Channel, the tidal streams cause so much turbulence and vertical mixing that the temperature of the surface is almost the same as that of the water below. In other regions, however, the gradient, dependent largely upon wave motion, is so variable, that surface temperatures alone are of little use in comparing the temperature distribution of the main water mass from one year to another during the summer months.

Wegemann ((9) 1922) has reviewed data concerning the diurnal variation of surface temperatures obtained for the most part during the cruise of the *Gazelle*; these bring out the influence of two factors—whether the sky is cloudy or clear, and the strength of the wind and consequent turbulence of the water. He quotes mean values for the diurnal variation based on observations by Merz in the Gulf of Trieste, at the surface, and mentions that variations were perceptible down to a depth of 15 metres.

Mean daily surface variation,	winter	0·71° C.
,, ,,	summer	0·89°
,, ,,	bright summer day ...	1·27°
,, ,,	cloudy ,, ,, ...	0·75°

INFLUENCE OF CURRENTS

In a vertical section north and south through the Atlantic, the influence of the movement of the water masses is apparent. In the equatorial region, surface water increases in salinity owing to evaporation but decreases in density owing to rise in temperature

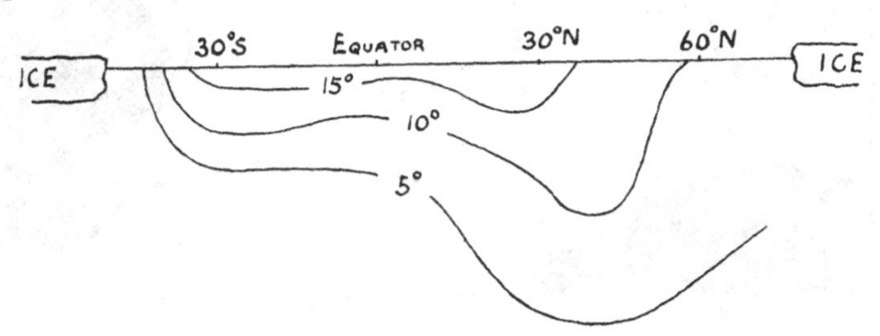

Fig. 48. Diagrammatic section through the Atlantic Ocean showing distribution of temperature with depth (Schott).

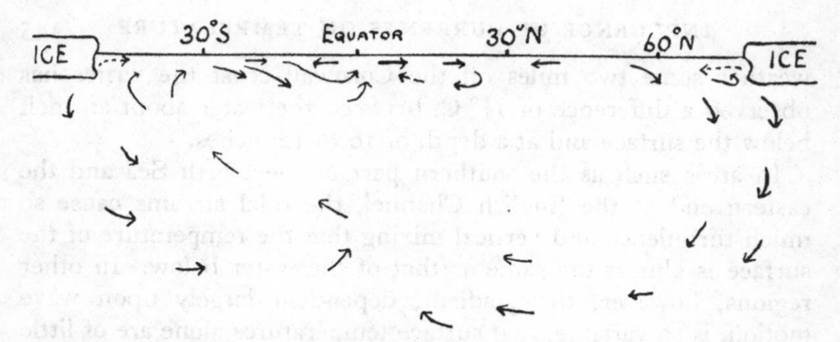

Fig. 49. Diagrammatic section through the Atlantic Ocean showing drift of deep water.

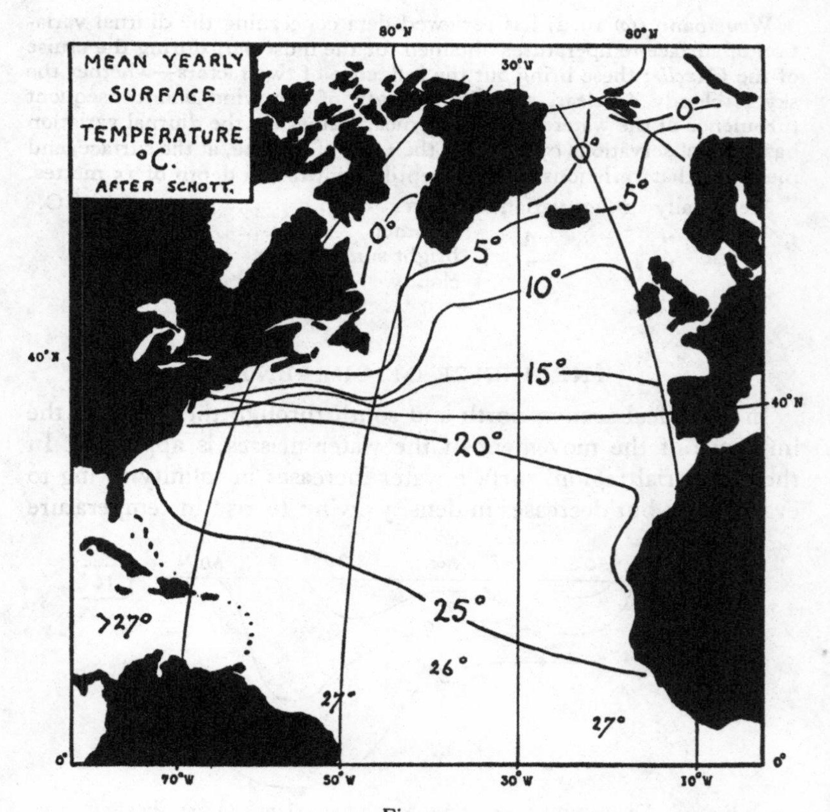

Fig. 50.

and passes westwards in the Gulf Stream, aided by the trade winds. Colder less saline water which has travelled along the bottom from Arctic regions wells up from below to take its place. Part of the warm saline Gulf Stream water passes into the Sargasso current and forms a pool owing to the anticyclonic wind. In the polar

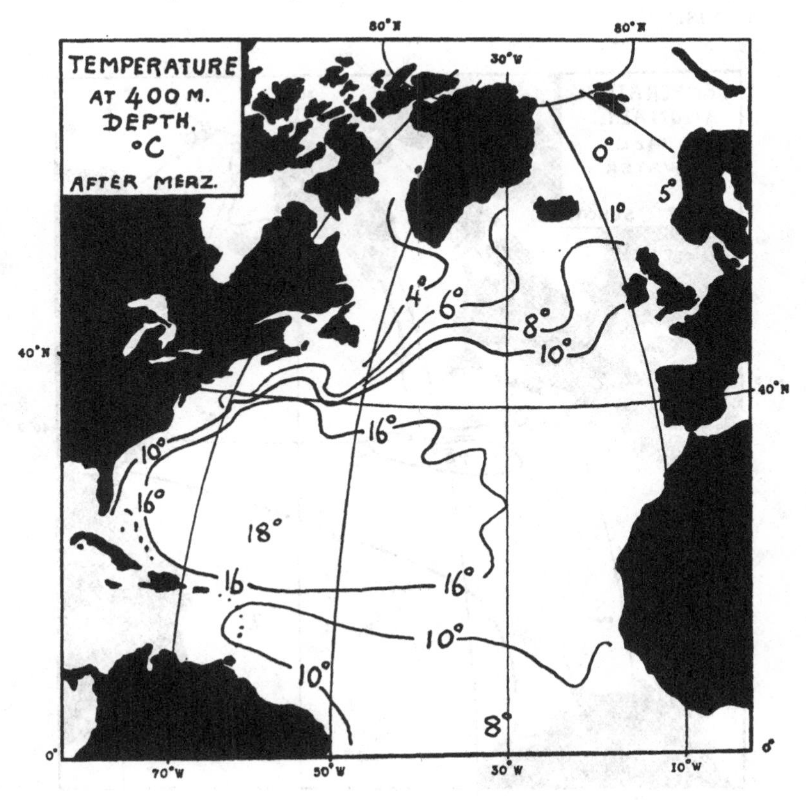

Fig. 51.

regions saline water from the Atlantic stream replaces the low salinity ice water which flows as an upper stratum southwards along the Greenland and North American coasts; part of this incoming saline water is cooled and, sinking gradually, moves as a bottom drift back to the equatorial regions where it once more finds its way to the surface.

As latitude increases going north or south the total of solar radiation received throughout the whole year decreases. Hence if there were no ocean currents the mean yearly isotherms in the sea would run roughly parallel with the lines of latitude. Their deflection from the parallels of latitude is mainly due to ocean currents.

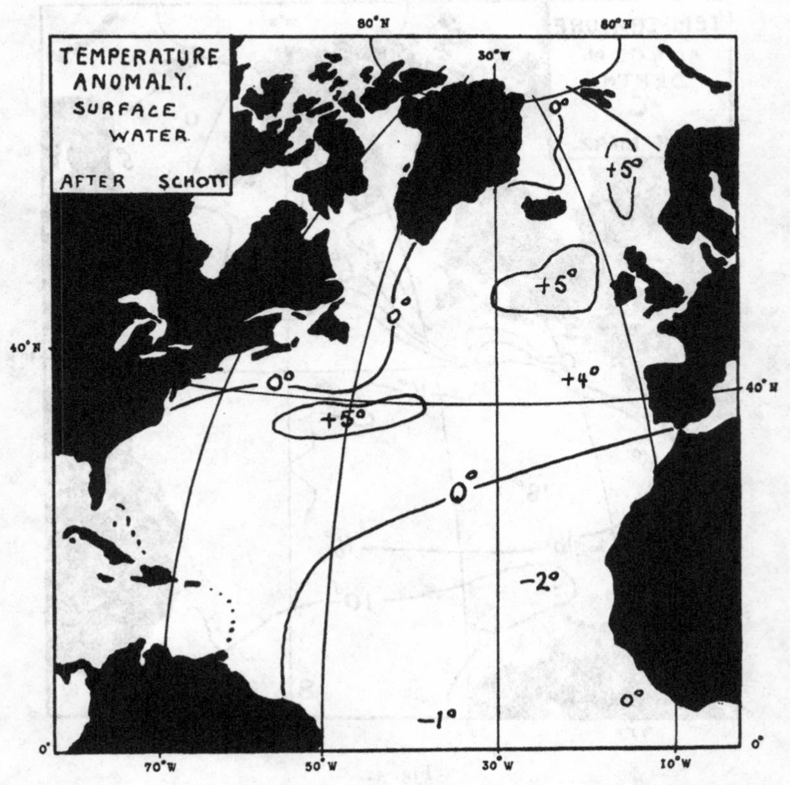

Fig. 52. Departure of the mean yearly surface temperature of the N. Atlantic from the mean temperature calculated for each latitude from the temperatures in the Atlantic, Pacific and Indian Oceans (Schott).

The influence of the currents in the upper layers of the North Atlantic is reflected in the horizontal temperature distribution, as may be seen on comparing Figs. 50 and 51 with the rough map of the currents on p. 115.

The distribution at a depth of 400 metres shows the effect of the north-going Atlantic Stream, and the probable influence upon its temperature of water welling out of the Mediterranean during the summer. Nansen traced the latter as far north as Ireland. The influence of a marked flow of this Atlantic Stream water into the English Channel in the autumn of 1921 has been mentioned in the general discussion of ocean currents.

Schott (4) has calculated the mean sea surface temperature of all three oceans (Pacific, Indian and Atlantic) for each zone of latitude, and has plotted the departure from these values of the mean yearly temperature of the Atlantic. This is shown for the North Atlantic in Fig. 52 and brings out very clearly the magnitude of the effect of ocean currents. (Compare map of the currents, p. 115.)

ADJOINING OR PARTIALLY ENCLOSED SEAS

Where an adjoining sea is separated from one of the great oceans by a submerged ridge, the deep water has almost the same temperature from ridge level to bottom. It has no connection with the deep water of the adjacent ocean, lying below the level of the ridge.

The basin of the partially enclosed sea is filled with the densest water which has access to it.

In the Mediterranean during summer the upper layers, heated and concentrated by evaporation, sink and run out through the Straits of Gibraltar as a subsurface current, less saline and lighter water from the Atlantic taking their place and running in on the surface. This inflowing water mixes with the deep layers in winter when nearly isothermal conditions exist; convection currents refresh the deep water annually from above.

In the Baltic, on the other hand, a different type of circulation takes place. Light low salinity water runs out as a surface current, its place being taken by heavier water of greater salinity from the North Sea, which runs in as a bottom current through the Skager-Rack. The bottom of the basin of the Baltic remains full of more saline water, into which the convection currents set up by winter cooling of the surface cannot penetrate since, although warmer, it is of greater density. It is refreshed by the inflow of heavy water from the North Sea, strongest in the autumn and early winter, but not to a sufficient extent to keep it fully oxygenated.

Table XLI. Baltic Sea. July 1922.
Lat. 57° 24′ N., long. 19° 52′ E.

Depth (metres)	Temp. ° C.	Density	Oxygen, c.c. per litre
0	15·52	1·00418	6·78
20	10·60	1·00537	7·64
40	2·95	1·00614	8·74
60	2·20	1·00637	7·86
80	4·05	1·00835	3·72
100	4·55	1·00870	3·62
150	4·55	1·01004	3·72
200	4·60	1·01021	2·45

The light surface water which runs out from the Baltic through the Kattegat in the summer months is a warm layer, similar to the surface layers shown in the above table. The inflowing salt undercurrent varies in strength not only with the seasons, but fluctuates from year to year; this inflow is colder than the water running out on the surface, and the greater the strength of the inflow the more mixes with the surface water, raising its salinity and lowering its temperature. Johansen and Jensen ((5) 1926) have recently shown that not only the surface salinity and surface temperature in June follow each other closely year by year, but also the air temperature in Copenhagen and Göteborg. Going one step further, the surface salinity in June follows the bottom salinity at the entrance to the Kattegat in April and May, since a greater salinity of the bottom water here is tantamount to a greater flow through the Kattegat into the Baltic. Hence a high salinity in the bottom water of the northern Kattegat in the spring is normally followed by a low air temperature at Copenhagen and Denmark in June and July.

There is yet a further correlation. Johansen ((6) 1925) has suggested that the annual catches of mackerel vary with the temperature of the upper layers, being greater during those years when the upper layers are cooler and *vice versa*. Information of the salinity of the bottom water in April and May, as indicating the probable temperature of the upper layers in the summer and forecasting the summer catch of mackerel, is now to be published in the *Dansk Fiskeritidende*.

In the Norwegian Sea and Arctic Ocean, separated from the Atlantic by a submarine ridge extending from Iceland to Scotland, the water below the ridge level is derived from the north. The section in Fig. 53 shows the distribution of temperature with depth.

In the Arctic Ocean and Mediterranean the temperature rises slightly between ridge level and the bottom. This increase in temperature appears to be due, in part at least, to adiabatic heating of the dense water as it was compressed.

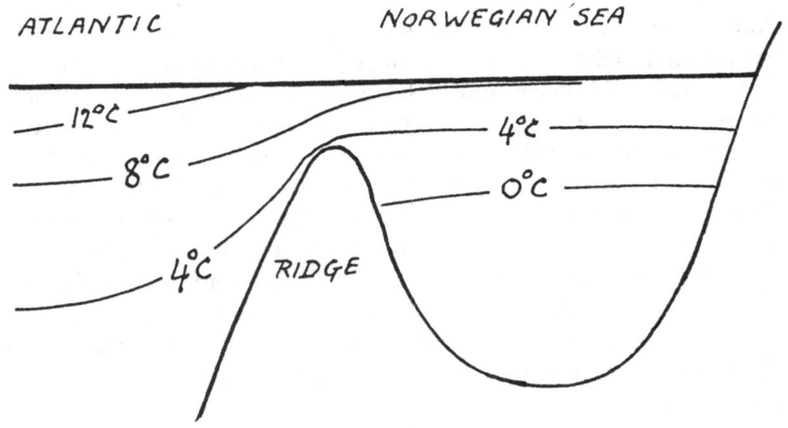

Fig. 53. Diagrammatic section showing the distribution of temperature with depth, between the open Atlantic and the coast of Norway across the ridge extending from Scotland to Iceland.

The Black Sea is an instance of the occurrence of stagnant deep water, practically isothermal and unaffected by seasonal variations below 150 metres; this water is depleted of oxygen and contains sulphuretted hydrogen from 0·5 c.c. per litre at 150 metres to 6·8 c.c. per litre at the bottom. Towards the coasts the isotherms and isosteres dip downwards, and the oxygen-containing upper layer, in which vertical mixing takes place owing to convection currents, is thicker. Since the sea is tideless this greater depth of oxygen-containing water near the shores must be due to the action of off-shore and on-shore winds.

FLUCTUATIONS OF TEMPERATURE

A very limited number of hydrographical expeditions into the oceans and numerous observations of the surface conditions have built up our present knowledge of the *average* state of the ocean currents, and have made it apparent that considerable fluctuations occur, both seasonal and of longer duration.

Fluctuations in the flow of the Atlantic Stream affect the temperature of the water around our coasts and of the upper layers in the Norwegian Sea. Early investigations by Pettersson, from scanty data, showed a decided relation between sea temperature close to the Norwegian coast, the air temperature, and various seasonal occurrences such as the time of flowering of particular plants in Norway. Later Helland-Hansen and Nansen ((8) 1909) found a correlation for some years between the surface temperature of the sea to the west of Norway and the yearly growth of pine. They calculated the quantity of heat brought into the Norwegian Sea by the Atlantic Stream, and found evidence of a close relation between this quantity in May from year to year and the mean air temperature over Norway during the following winter. This relation held for the years during which the observations were made (1900–1905).

A certain amount of evidence has been obtained that these fluctuations occur at more or less regular periods. The amount of Atlantic water flowing into the Norwegian Sea and raising the surface temperature was found to be less every second year from 1874 to 1904 with few exceptions. A similar periodicity occurred in the surface temperature of the North Sea in November and on the route Humber-Elbe. The most striking conjecture so far advanced is the theory by Pettersson that when the horizontal pull due to attraction towards the moon is at a maximum, the flux of the Gulf Stream is also at its highest. This occurs every 18½ years when the moon's declination is least. From the data available it does appear that the northerly flow of high salinity water was at a maximum in 1904–1905 and 1921, when this condition was fulfilled. The possibility of regular periodicity in the ocean currents has led to interesting speculation, particularly by Pettersson who has collected and analysed an extensive body of evidence.

It is apparent from a consideration of the effect of meteorological conditions upon the gain and loss of heat by the sea, that annual temperature fluctuations will occur also in areas not subject to incursions of water from other latitudes, although in these cases the fluctuations are likely to be less. If a particular stratum of water is under consideration, changes in the amount of vertical mixing which takes place will play a leading rôle in these fluctuations.

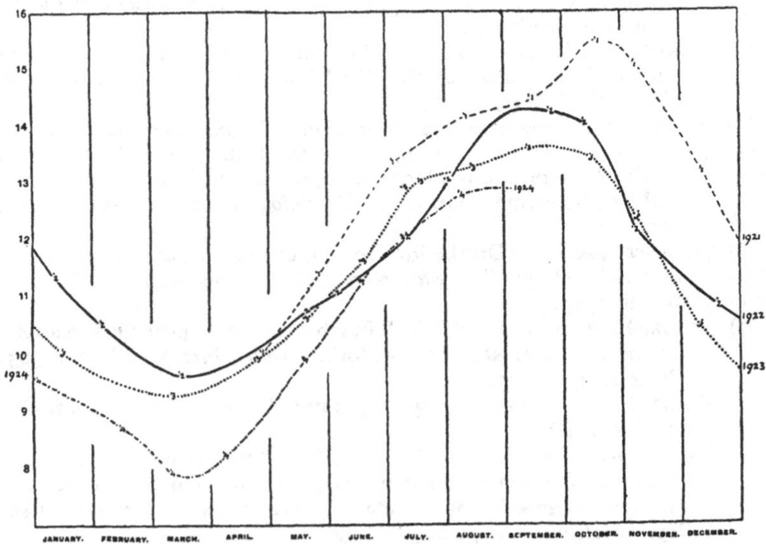

Fig. 54. Temperature in degrees centigrade of a column of water at lat. 50° 02′ N., long. 4° 22′ W., from top to bottom, as if completely mixed, for the years 1921, 1922, 1923, 1924.

By determining the mean temperature of the whole column of water from top to bottom—the integral temperature—the gain or loss of heat at a particular position over a period of time may be found. Fluctuations from one year to another are due to fluctuations in the various factors already dealt with, including, of course, movements of the water masses. Fig. 54 shows the effect of the inflow of Atlantic water in 1921 upon temperature in the English Channel very clearly, and also shows that during the following two years, when there was probably little movement of the water masses, the annual variations were not identical ((2c) 1925).

BIBLIOGRAPHY

(1) ÅNGSTRÖM, A. *Journ. Roy. Met. Soc.* **50**, 121. 1924.

(2 *a*) LÜTGENS, R. "Die Verdunstung auf dem Meere." *Ann. Hydrog. u. Mar. Met.* **39**, 410. 1911.

(2 *b*) SCHMIDT, W. "Strahlung und Verdunstung an freien Wasserflächen u.s.w." *Ann. Hydrog. u. Mar. Met.* **43**, 111–24. 1915.

(2 *c*) HARVEY, H. W. "Evaporation and Sea Temperature." *Journ. Mar. Biol. Assoc.* **13**, 678–92. 1925.

(2 *d*) MCEWEN, G. F. "Ocean Temperatures." *Semicentennial Public. Univ. California.* 1918.

(3) ATKINS, W. R. G. "On the Thermal Stratification of Sea Water and its Importance for the Algal Plankton." *Journ. Mar. Biol. Assoc.* **13**, 693–99. 1925.

(4) SCHOTT, G. *Geographie des Atlantischen Ozeans.* Hamburg, 1926.

(5) JOHANSEN and JENSEN. "Remarks on the Influence of the Currents in the Waters about Denmark upon the Climate of Denmark and Neighbouring Countries." *Physiological Papers.* Copenhagen, 1926.

(6) JOHANSEN, A. C. "On the influence of currents upon the frequency of mackerel, etc." *Medd. Komm. f. Havunders. Serie Fisk.* **7**, No. 8. 1925.

(7) BIRGE, E. A. and JUDAY, C. "Further Limnological Observations on the Finger Lakes of New York." *Bull. Bureau Fish.* No. 905. Washington, 1921.

BIRGE, E. A. "Limnological Apparatus." *Trans. Wisc. Acad. Sci. A. and L.* **20**.

(8) HELLAND-HANSEN, B. and NANSEN, F. "Die jährlichen Schwankungen der Wassermassen im norweg. Nordmeer in ihrer Beziehung zu den Schwank. der meteor. Verhaltniss." *Internat. Rev. Hydrobiol. u. Hydrog.* **2**, 337. 1909.

(9) WEGEMANN, G. "Der tägliche Gang der Temperatur der Meere und seine monatliche Veränderlichkeit." *Wiss. Meeresuntersuchungen*, **19**, 51. Kiel, 1922.

Chapter V

COLOUR AND ILLUMINATION OF SEA WATER

Two distinct changes take place when a pencil of white light, composed of rays of different wave lengths which constitute daylight, passes through water containing no suspended particles. A certain number of the rays are *absorbed* by the water and converted into heat; the red and orange rays are more rapidly absorbed than the green, blue and violet. In sea water these longer wave lengths at the red end of the spectrum are absorbed even more rapidly than in pure water. The shorter ultra-violet rays, such as possess germicidal power, are rapidly absorbed like the red rays. In this connection the recent work of Duclaux and Jeantet is of interest, for they find that traces of ammonia, proteins and nitrates in solution in pure water greatly reduce its transparency to short wave length ultra-violet light, whereas dissolved salts in general have little effect. They find that such ultra-violet light is rapidly absorbed by well-water and by sea water. In addition to this selective absorption the light rays undergo *scattering* by the molecules of the water, being deflected from their straight path, some even being returned along their original path. The short wave length rays are scattered to a much greater extent than the longer, and for this reason the path of a pencil of rays passing through optically pure water appears blue, and the colour of a deep pure water, when viewed from above, is blue.

The rate of absorption of light of varying wave length has been investigated by several observers. It is generally expressed as the coefficient of absorption α, where

$$I_h = I_0 \epsilon^{-\alpha h}$$

and I_h = intensity of the light after travelling through a distance h, usually expressed in metres;

I_0 = intensity of the light which penetrates the surface;

ϵ = base of natural logarithms, 2·7183.

This coefficient α is the reciprocal value of the distance the light passes through the water whereby its intensity is reduced to $\frac{1}{2\cdot7183}$ of its value at entrance. In the sea where plankton organisms are not evenly distributed with depth, the value of α may change slightly on passing from the upper to the lower layers. In optically pure water it is constant and the fraction of the light lost in its passage through each consecutive metre is the same. Very considerable discrepancies occur between the values of α for pure water obtained by different observers, probably owing to the difficulty of obtaining water entirely free from minute suspended particles. Table XLII shows typical values.

Table XLII.

Colour	Wave length in $\mu\mu$ (millionths of a millimetre)	Absorption coefficient per metre
Violet	415	·03
Indigo	450	·02
Blue	470	·02
Blue green	490	·02
Green	530	·02
Yellow-green	550	·03
Yellow	590	·1
Orange-yellow	615	·2 } Within the
Red	660	·25 } absorption band

0·001 millimetre = μ = 1000 $\mu\mu$ = 10 Å.

In nature 'optically pure' water never occurs, suspended inorganic matter and myriads of minute living organisms being always present; these reflect light rays passing through the water from their surfaces, besides absorbing a portion of the light. The greater the number of suspended particles and organisms, the greater is the quantity of white or yellowish light of mixed wave lengths which, together with the blue from molecular scattering, is seen as a blue-green or green colour when looking down into water lit from above.

The transparency of the sea may be estimated by noting the depth at which a white disc (Secchi's disc) lowered vertically into the water just becomes invisible. The depth is a measure of the transparency of the water and is not materially affected by the intensity of light at the surface within fairly wide limits. The disc becomes invisible at a depth where the same fraction of the light entering the water has been absorbed; thus the depth noted is a rough inverse measure of the effective amount of suspended matter. During the cruise of the *Deutschland* to the Weddell Sea a number of observations were made of the colour of the sea when looking down from the ship, and the depth at which such a disc became invisible, with the following result:

Table XLIII.

Colour of the sea as viewed from above	Mean depth at which Secchi's disc became invisible
Deep blue ...	35 metres
Blue	27 ,,
Blue-green ...	18 ,,
Green-blue ...	12 ,,
Green	9 ,,

In the very deep blue Sargasso Sea the disc was visible to 66 metres, and in the blue water between pack-ice in the Antarctic to a mean depth of 50 metres.

A green colour as viewed from above is characteristic of water containing many individual organisms, or particles of suspended matter carried out from the land, while a deep blue colour is a characteristic of the desert areas of the oceans.

Passing next to the character of the light which penetrates the sea and would be seen by a human eye at different depths, variants between two conditions may occur. In 'transparent' water, where a white disc can be seen from above to a considerable depth and where the water contains little in suspension and is characterised by a deep blue colour, the observer will experience on sinking a change from white light to green and finally to blue-green light, because the wave lengths at the red and violet ends of the spectrum are more rapidly absorbed than around the blue region. The intensity of the light will rapidly diminish at first and then more slowly, since the intensity will decrease by the same fraction for each equal increment in depth (see Fig. 55). In a coastal water, or one rich in plankton life, where a disc can only be seen down to a limited depth and where the water viewed from above appears green, the same change from white to greenish to green light would be experienced, and owing to the particles shutting out light rays of all kinds the intensity of light will diminish more rapidly with depth than in the former case. At the same time the change in colour with increasing depth from green to blue-green will be less rapid. The suspended particles, in addition to reflecting and absorbing rays of all wave lengths, scatter the short wave length blue rays to a much greater extent than the longer yellow and green rays. Hence at the depth where the light from above appears blue-green in transparent water, the proportion of blue to longer wave length rays will be less in water rich in suspended particles. In such a case the apparent absorption of the blue rays is increased more than the apparent absorption of the green rays.[1]

The foregoing summarises the change in *quality* of white light as it penetrates the sea. The figure 55 roughly illustrates the sensitivity of the eye to various wave lengths; as the intensity

[1] Pietenpol found the coefficients of absorption for light of different wave lengths through filtered and through unfiltered lake water containing many suspended particles and traces of colouring matter. From his results the percentage of yellow and of blue light which passes through 1 metre of each water was calculated with the following results:

Table XLIV.

Wave length $\mu\mu$	Light entering water	Percentage transmitted through 1 metre of filtered lake water	Percentage transmitted through 1 metre of unfiltered lake water
558 (yellow)	100	83	$31\frac{1}{2}$
470 (blue)	100	$72\frac{1}{2}$	17

The proportion of blue to yellow which penetrates the unfiltered lake water is very much less than the proportion which penetrates filtered water. It should be noted that Pietenpol drew an exactly opposite conclusion from his observations, which are expressed in terms of the coefficient of absorption α; however, on calculating the percentages of light transmitted through 1 metre of the filtered and unfiltered water a smaller *proportion* $\left(\frac{17}{31 \cdot 5}\right)$ of blue to yellow light is found to penetrate the unfiltered water than penetrates the filtered water $\left(\frac{72 \cdot 5}{83}\right)$.

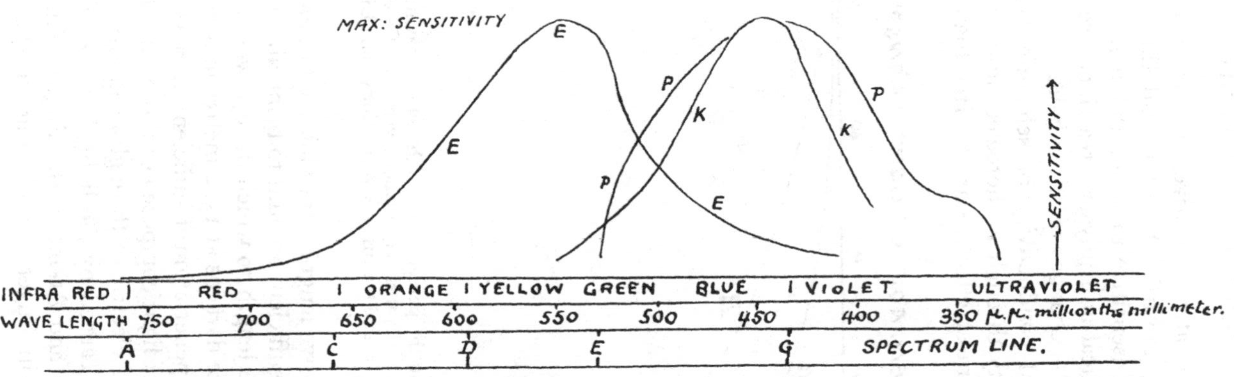

Fig. 55. Sensitivity of the human eye E, the potassium hydride photoelectric cell K, and an ordinary photographic plate P to light of different wave lengths in the solar spectrum.

becomes low the relative sensitiveness to the colours changes somewhat, shifting towards the blue, and the range decreases, the eye becoming insensitive to violet and red.

Numerous chemical and physical reactions are brought about by light, such as the changes of silver halides in a photographic plate, the electrical conductivity of selenium crystals and the passage of electricity through a photoelectric cell: here the wave lengths which are most effective are shorter than in the case of the eye.

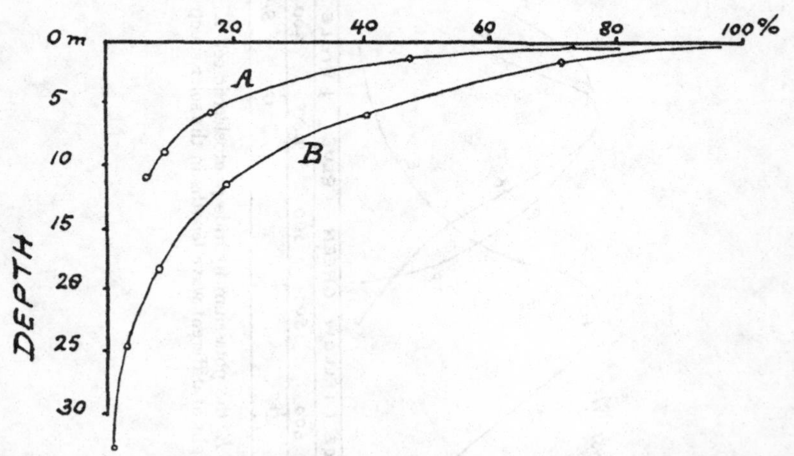

Fig. 56. Curves showing the percentage of light which, having entered the sea, is transmitted to various depths. *A*. Inshore water in Cawsand Bay, depth 12 metres. *B*. English Channel ten miles S.W. of Eddystone Lighthouse. (Poole and Atkins.)

Violet and long wave length ultra-violet light penetrates to considerable depths in sufficient amount to bring about such changes. These are of a wave length to which the eye would be insensitive. A photographic plate darkens at 1500 metres in ocean water in the tropics owing to the penetration of sufficient short wave length light.

The penetration of light composed of those wave lengths, which affect the passage of electricity through a potassium hydride cell owing to ionisation caused by the light, has been investigated by Poole and by Poole and Atkins in the English Channel, and by Shelford and Gail in Puget Sound. Fig. 56 shows the rate of decrease of the light with depth in the clear water of the English Channel and in Cawsand Bay, where the water contains many suspended particles. Only some 5 per cent. of the light striking the surface of the sea was found to be reflected with a smooth sea and grey sky, whereas with breaking waves as much as 31 % was reflected.

Grein observed the actinic strength of red, green, blue and ultra-violet rays in clear water in the Mediterranean, where a white disc was visible to a depth of over 30 metres, by screening off the unwanted rays by filters from a Wratten photographic plate let down into the sea. He expressed the actinic value in terms of the depth of colour given by the plate when exposed to a standard Hefner lamp at 1 metre for 1 second. In these terms an intensity of blue light having slight effect upon the eye will register considerable candle power, the sensitiveness of the plate extending much further into the region of low wave length light than the sensitiveness of the human eye.

The following values are taken from his results and illustrate the fact that it is the blue and violet light which has the greatest *chemical* activity at a depth in a clear sea.

Table XLV.

Depth	Metre second candles			
	Red	Green	Blue	Ultra-violet
5 metres	26	—	—	—
50 ,,	—	495	3,950	19,500
100 ,,	—	52	150	1,250
At 1 p.m. sun's altitude 68° 8′				
5 metres	2·8	—	—	—
50 ,,	—	14·2	6·6	190
100 ,,	—	1·5	4·5	47·4
At 7 p.m. sun's altitude 0°				

Grein noticed that the quality of the light falling upon the sea altered in the afternoon and evening, when the proportion of long wave length light to the shorter wave length increases. The most noticeable case is that of a red sunset when blue and ultra-violet light is scattered by water or dust particles in the atmosphere. Grein's results showed a greater proportion of green to ultra-violet in the sea at 50 and 100 metres depth in the evening than in the morning due to turbidity of the atmosphere caused by evaporation from the sea during the heat of the day. In these observations quartz filters and Wratten 'speed' plates were used in measuring the intensity of the ultra-violet light, while Wratten panchromatic plates, sensitive to the red end of the spectrum, were used for red and yellow light.

Knudsen made use of an apparatus with direct vision spectroscope and glass lens which was lowered into the sea in Nyborg Fiord. The depth was 9 metres and plankton was abundant; the water evidently contained many suspended particles and organisms, since a white disc could only be seen to a depth of 4 metres. Under these conditions light of short wave lengths may be expected to undergo considerable scattering by the particles, and Knudsen's results support this expectation, for he found that green light (510 $\mu\mu$) penetrated furthest.

All the methods used so far have been subject to limitations in some respect, as pointed out by Klugh, whose review of the question from the standpoint of the plant physiologist is valuable.

The intensity of illumination and the quality of the light at any point below the surface of the sea depend not only upon the penetration of the light through the water and the scattering which it undergoes, but also upon the variations in intensity and quality of light falling upon the surface, upon the proportion reflected and upon the angle at which it meets the water. In the formula

$$I_h = I_0 \epsilon^{-ah}$$

h refers to the distance the rays travel through the water. Where they strike at an average angle A to the vertical, the formula becomes

$$I_d = I_0 \epsilon^{-a(h \operatorname{cosec} A \times \eta)},$$

where η is the index of refraction, approximately 1·333 for sea water, and h cosec $A \times \eta$ is the distance the rays travel through the water before reaching the depth d in question.

The rays in the water never make an angle with the vertical greater than $48\frac{1}{2}°$, which is the critical angle.

The height of the sun and the length of day are periodic variations. On a sunshiny day part of the illumination reaching the surface is direct sun's rays and part light from the hemisphere of sky, the amounts varying more or less rapidly with the passage of clouds, which, passing across the sun, reduce the direct sunshine radiation, and passing near the sun increase materially the light from the sky. As already mentioned the quality of the light reaching the sea's surface varies, in the relative proportion of rays of different wave length composing it, with the atmospheric conditions.

BIBLIOGRAPHY

SHELFORD and GAIL. "A Study of Light Penetration into Sea Water made with the Kunz Photoelectric Cell." *Publ. Puget Sound Biological Station*, **3**, 141. 1922.

PIETENPOL, W. "Selective Absorption in the Visible Spectrum of Wisconsin Lake Waters." *Trans. Wisc. Acad. of Sciences*, **19**, Part I. 1918.

KNUDSEN, M. "Penetration of Light into the Sea." *Pub. de Circonstance*, **76**. Copenhagen, 1922.

GREIN, K. "Untersuchungen über die Absorption des Lichts im Seewasser." *Ann. de l'Inst. Océanographique de Monaco*. 1914.

POOLE, H. H. "On the Photoelectric Measurement of Submarine Illumination." *Sci. Proc. Roy. Dublin Soc.* **18**, 99. 1925.

POOLE, H. H. and ATKINS, W. R. G. "On the Penetration of Light into Sea Water." *Journ. Mar. Biol. Assoc.* **14**, 177. 1926.

DUCLAUX and JEANTET. "Transparence des Eaux Naturelles aux Rayons Ultraviolets." *C.R.* **181**, 630. 1925.

KLUGH, A. B. "Ecological Photometry and a New Instrument for Measuring Light." *Ecology*, **6**, 203–37. 1925.

THOULET, J. "Étude sur la Transparence et la Couleur des Eaux de Mer." *Résultats des Campagnes Scientifiques*, Fasc. 24, 113–34. Monaco, 1905.

ATKINS, W. R. G. "A Quantitative Consideration of some Factors concerned in Plant Growth in Water." *Journ. du Cons. Internat.* **1**, 99–126. 1926.

Chapter VI

CHEMICAL AND PHYSICAL FACTORS CONTROLLING THE DENSITY OF POPULATION

The foregoing chapters deal with our present miscellaneous and often sporadic knowledge of the physical and chemical conditions which control the amount of plant and animal life supported by various areas of the ocean waters. From this knowledge certain firmly interrelated principles emerge, upon which it now becomes possible to build up a coherent theory—a theory which will, without doubt, require modification as research progresses, but nevertheless valuable as it provides a working hypothesis upon which to base future investigations relating to the productivity of various areas. A number of factors potent in their effect upon marine life—temperature, currents, light, and nutrient salts for plant life—and their close relation the one with the other, have been considered. Research may disclose other factors which enter into the mosaic of conditions controlling the population: at all events we are dealing with *living organisms* which are subject to the resultant of so many physical laws that their behaviour is extremely complex. Our knowledge is partly based on facts ascertained for particular species under particular conditions, and, however probable, such facts are not necessarily true under all the varied conditions which occur in the sea; consequently any theory at present can only be of a provisional nature and indicate broad generalisations.

GROWTH AND CONSUMPTION OF ALGAE

At the outset of any consideration of fertility—the amount of living plant and animal tissue per unit area of the sea which various areas are capable of supporting—it is clear that under circumstances such as occur in nature, where the growth of plants is neither regular in rate nor regular in the proportion of the species which constitute the flora, a considerable proportion must die and decay at times of glut. Then more is produced than is necessary to feed the animals present. This also happens when a sudden burst of growth takes place of some one species (such as *Phaeocystis*) which is not a suitable food for the majority of plankton-feeding animals. Upon the proportion of plant growth which dies and decays, and the proportion which goes to energise animal organisms, depend in part the total fertility of any area. It cannot be assumed that where vegetable carbohydrate and protein food exist, there will be an animal population to eat a *fixed* proportion of it. Besides a supply of phytoplankton suitably spaced in time, a further condition necessary for maximum population is that the energy of plant life passed on to the phytoplankton-feeding animals should, before they die, be handed on to carnivorous animals. If the phytoplankton-feeding animals succumb to a natural death, part at least of their energy will be dissipated during the breakdown by autolysis of their body substances. Probably both these conditions are not so important as appears at first sight, since the corpses of most dead organisms form the food of numerous protozoa and bacteria which themselves provide the nutriment of other animals. In this connection it is interesting that, even in the great depths of the oceans, a fauna exists on the bottom dependent for its energy upon the rain of *dead* organisms from above.

Lohmann has endeavoured to calculate the relation between the augmentation of the algae and their consumption by marine animals in Kiel Bay, where he studied the quantities of plankton for a whole year and calculated the volume of the various groups in the different water masses at all seasons. On an average plants made up 56 % and animals 44 % of the total plankton: in the winter months the plants were outnumbered, being scarcely one-third of the total plankton from December to February: in the summer they predominated and made up sometimes as much as three-quarters of the whole. By assuming that the phytoplankton increases in volume by 30 % daily, and that this increase can be consumed by the animals without harm to the plant aggregate, and by assuming that protozoa require daily one-half of their own volume of food and other animals one-tenth, Lohmann has compiled a table showing the surplus or deficiency of plant food each month for the animal population. During the summer months this table shows a considerable surplus, several times the calculated requirements of the animals, but from November to the end of February the surplus is not large: in February there was actually a deficiency. However, owing to the slower rate of metabolism during the winter, the animals may require proportionately less food.

Table XLVI.

	Total phyto-plankton. Cu. mm. per cu. m.	Total animal plankton. Cu. mm. per cu. m.	Calculated surplus or deficiency of phyto-plankton per cu. m. available for nutriment of animal plankton
August, 1905	1095	464	+290
September	822	682	+190
October	425	457	+ 85
November	262	371	+ 45
December	99	244	+ 10
January, 1906	54	158	+ 12
February	31	145	− 8
March	102	186	+ 6
April	390	134	+110
May	462	495	+ 85
June	623	228	+160
July	519	320	+125
August	505	321	+117

In the open sea of temperate regions generally, phytoplankton is often sparse at times between the onset of growth in spring and the late summer or autumn outburst. Since animal plankton is numerous between these dates, it is possible that there may occur from time to time a shortage of vegetable nutriment. In the shallow waters of Kiel Bay, where Lohmann based his conclusions that the summer period is one of surplus plant food, a maximum of phytoplankton occurred in the late summer and there was no marked spring outburst followed by a decline, as is usual in most open sea areas.

The occurrence of a shortage at times during the summer receives a certain amount of confirmation from observed changes in hydrogen ion concentration and dissolved oxygen. If the production of phytoplankton falls below the requirements of animal life the amount of CO_2 set free in respiration tends to increase above the amount utilised in photosynthesis. In consequence the hydrogen ion concentration rises. The reverse occurs when the rate of photosynthesis exceeds the rate at which CO_2 is set free owing to respiration by the whole community, as happens in the spring. Measurements of hydrogen ion concentration in the English

Channel (2) indicated that in June and July 1922 respiration exceeded photosynthesis, while the 1923 observations gave no such indication. In the former year sunshine was at a high maximum in May, while in the latter year it was more evenly distributed throughout the summer. Marshall and Orr's measurements of hydrogen ion concentration and dissolved oxygen show that respiration exceeds photosynthesis during the periods between outbursts of diatom growth in the Clyde sea area during the summer months.

It is a temptation to compare the production of life in the sea with the productivity of pasture in an agricultural country; although fundamentally comparable in that both are dependent upon plant life, in the sea carnivorous and partly carnivorous animals flourish unchecked, while in cultivated countries they are treated as vermin, man retaining almost exclusively for himself this particular diet. Comparison with the productivity of a wild and uncultivated country is more reasonable, yet even in this case there is the difference that land plants and animals are restricted more or less to the surface while in the sea they occur throughout the depth.

CYCLE OF LIFE IN THE SEA

A consideration of the chain of marine organisms which make up the flora and fauna brings out very clearly that certain factors control the potential fertility. The chain may be summarised as follows:

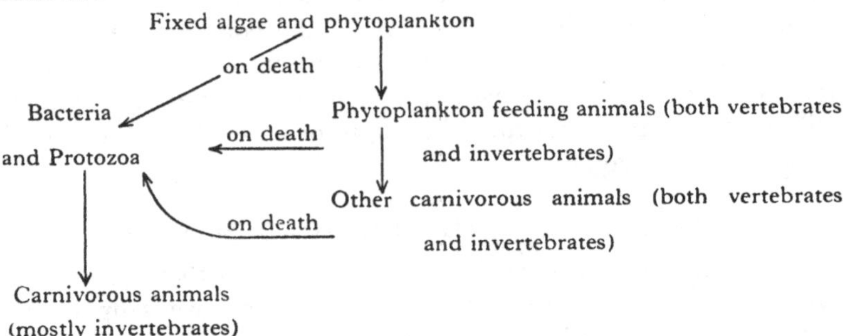

This leads directly to the next step showing that the cycle in the sea forms a nearly closed system.

The energy, absorbed from sunlight and used in photosynthesis, is in equal amount set free as heat by the living organisms and in the final breakdown of their corpses.

Since there is an excess supply of the requirements for photosynthesis with the exception of phosphate and nitrate, the closed system can be written as follows:

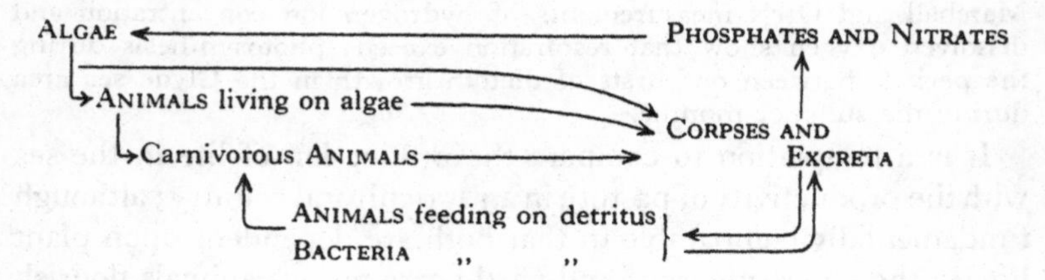

This schema shows that the fertility of an ocean will depend for the most part upon two factors, namely the length of time taken by the corpses of marine organisms and excreta to decay, and the length of time taken by the phosphates and nitrates so formed to come again within the range of algal growth. Where the corpses fall in deep water, well below the light intensity necessary for photosynthesis, and where there is no vertical mixing of the water or deep currents to take the phosphates and nitrates to lesser depths, they are likely to remain lost for many years to the cycle of life owing to the long period which must elapse after the salts are re-formed from dead organisms and before they again reach the upper sunlit layers. The process of decay will be slower in the cold bottom water of a deep ocean than in the comparatively warm bottom water of shallow areas. In most places it is the time taken for the re-formed salts to be transported to the upper layers which will predominate, rather than the length of time taken by the dead organisms to decay.

UPWELLING AND VERTICAL MIXING

From these *a priori* considerations, it is evident that those areas, where deep water rich in nitrate and phosphate wells up to the surface, are likely to be abundantly stocked with life, both planktonic and otherwise, as was recognised by Nathansohn ((3) 1906). In the North Atlantic this occurs where the drift of deep water runs up shallow plateaux, or over submarine ridges as in the neighbourhood of the Faeroes and Iceland. Such an upwelling occurs along a coastline owing to the action of strong off-shore winds. The abundant life on the fringes of steep-to oceanic islands is doubtless due to this.

Another and perhaps more potent condition affecting fertility occurs where the surface water cools in winter giving rise to isothermal conditions and in consequence vertical mixing takes place

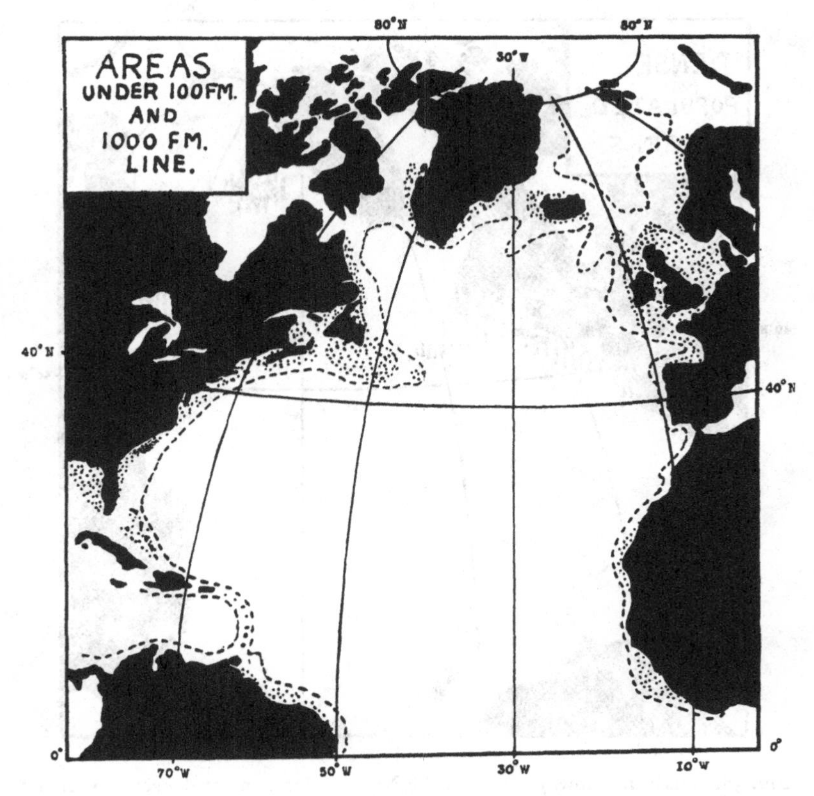

Fig. 57. Diagram showing areas in the North Atlantic with a depth below 100 fathoms. Dotted lines represent 1000 fathoms contour.

readily. Here nutrient salts formed from detritus on the bottom are brought each winter within the range of illumination. The development of phytoplankton in the early summer is consequently all the more abundant.

Figs. 57–59 illustrate these influences.

In the deep water of the open ocean below the limit which light sets to photosynthesis by the phytoplankton there is literally an enormous store of nitrate and phosphate. Judging by the quantities found below 100 metres at a position some 200 miles west of Portugal (pp. 41, 46) the dark depths of the three great oceans

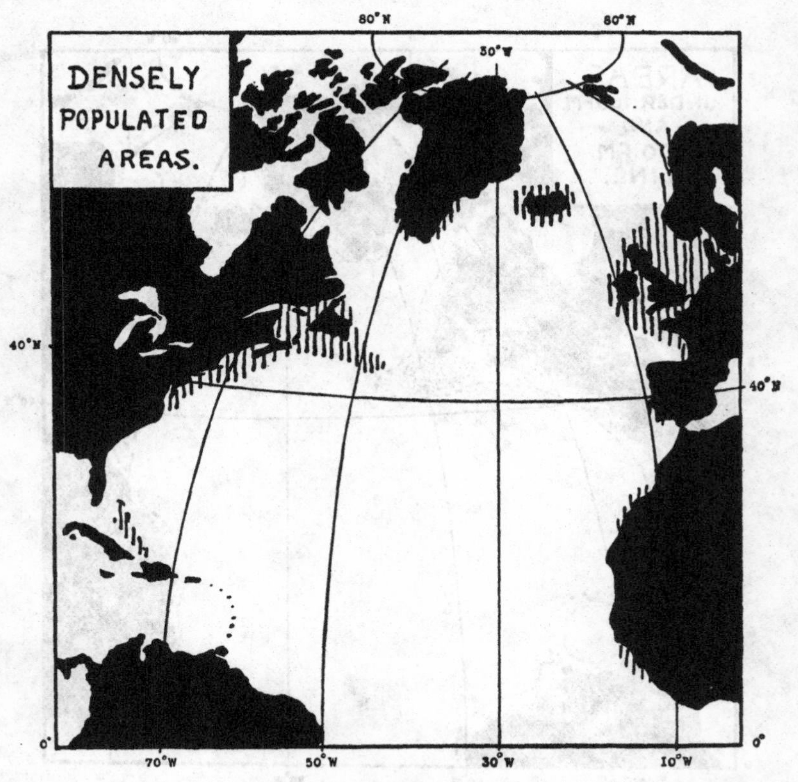

Fig. 58. Diagram showing areas in the North Atlantic most densely populated with marine animals, particularly fish.

contain some 250 thousand million tons of nitrate-nitrogen and 75 thousand million tons of phosphate reckoned as P_2O_5, which are lost to the cycle of events for an almost indefinite period. In low latitudes this is restrained from rising vertically to the surface owing to the lesser density of the warmer layers above, while diffusion of the salts through water is a vastly slow process. In higher latitudes the density usually increases less rapidly with depth and in consequence there is less restraint.

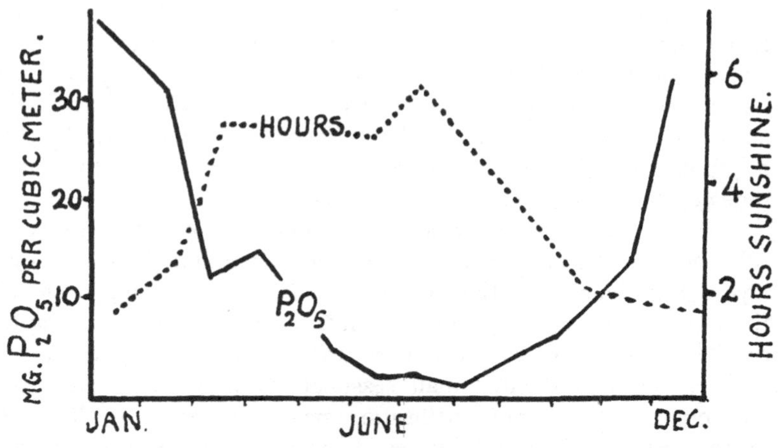

Fig. 59. Observed seasonal change in phosphate in solution in the surface water of the English Channel, 20 miles south-west of Plymouth breakwater, during 1924. The dotted line denotes the hours of sunshine per day.

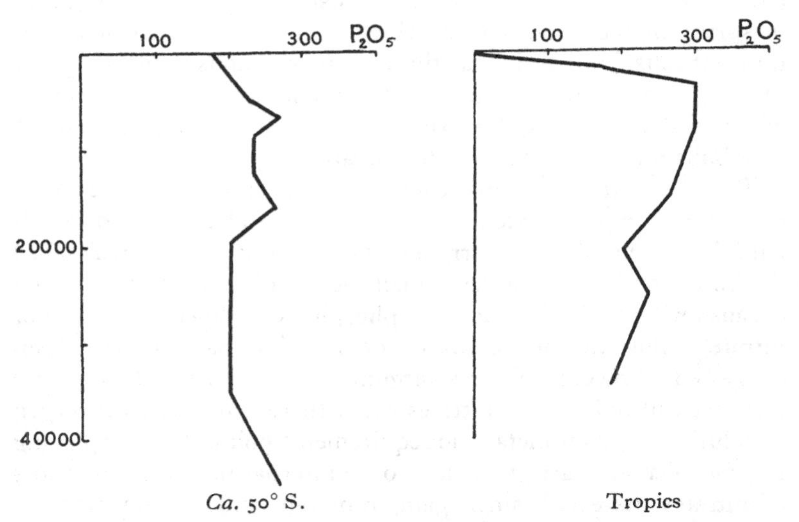

Fig. 60. Distribution of phosphate with depth in the water of the Antarctic and the tropical South Atlantic. Phosphate in milligrams P_2O_5 per cubic metre and depth in metres. (After Wattenberg.)

Figure 60 shows the amount of phosphate available for phytoplankton in the upper illuminated layers of the South Atlantic. In the tropics deep water rich in phosphate is debarred from rising through the discontinuity layer to mix with the less dense layers above, which have in consequence become depleted of phosphate by previous generations of phytoplankton. In high latitudes, on the other hand, a depletion of the upper layers by the phytoplankton does not occur because they are continually being renewed by strong convection currents.

The interval of time, which will elapse between the nutrient salts becoming available for use by phytoplankton in the upper layers and their utilisation, is likely to be a short one, provided the temperature and amount of sunshine are suitable for rapid growth of the flora. In this matter latitude and season play a part. In tropical waters with small seasonal variation in temperature and much sunshine, rapid utilisation may be expected at all times of the year; in higher latitudes during the winter months and short days with little sunshine, utilisation of the nutrient salts will be deferred. This actually happens during November, December and January around our coast, and a flush of diatom growth occurs in spring in the upper layers, which have become enriched with nutrient salts. In Arctic seas there is an enormous wealth of plant life in the short summer months after the long dark winter, during which the cold-blooded marine animals live at a slow rate of metabolism owing to the low temperature.

To what extent is this cycle 'almost closed' in the open oceans and to what extent modified in coastal areas into which land drainage and rivers carry a certain quantity of nutrient salts? There is known to be a very small loss of phosphate in the open oceans owing to the formation of phosphatic nodules, and a loss of nitrate owing to the action of denitrifying bacteria has been suggested; however Gran's argument that these denitrifying bacteria will only attack nitrates when there is insufficient oxygen in solution for their metabolic requirements points to their playing no conspicuous part (p. 12). To counterbalance any small loss of nitrates there is a small gain from rain and snow, which are stated to contain minute quantities of nitrous and nitric acids.

FERTILITY OF COASTAL AREAS

In the coastal areas the water is enriched by phosphates and nitrates carried down by rivers and streams. This undoubtedly permits a dense population, particularly by seaweeds in the estuaries of large rivers. During the summer months these salts are largely used up by the plant growth in the non-tidal and tidal parts of the rivers before reaching the open sea. Even during the winter when the rivers do carry down considerable quantities, their effect is not apparent beyond quite a limited distance from the shore in the cases which have been investigated, namely the rivers discharging into Plymouth Sound whose effect upon the open sea may be judged from Tables XLVII and XLVIII. There is evidence, however, that the waters of the Amazon bring down nutrient salts, or organic matter from which these are formed, and that these find their way seawards for a considerable distance. In the English Channel the nutrient salts which are regenerated annually from dead plants and animals eclipse the quantity added from the land drainage at a distance of more than two or three miles off shore.

Table XLVII. Milligrams of nitrate-nitrogen per cubic metre in the surface 10 metres at varying distance off shore.

In Plymouth Sound	Off W. end of break-water	6 miles S. 21° W. (true) from break-water	19 miles S. 21° W. (true) from break-water	56½ miles S. 21° W. (true) from break-water	Date
—	14	2	2	55	May 13, 1925
9	—	0	0	—	—
112	79	27	6	—	Oct. 1, 1925
176	380	76	67	—	Nov. 11, 1925
—	120	74	38	106	Dec. 11, 1925
>190	—	73	68	53	Mar. 11, 1926
135	—	110	75	—	April 10, 1926
11	—	5	4	30	May 17, 1926
24	—	5	<5	<5	July 8, 1926
20	—	8	10	—	Aug. 16, 1926
17	—	7	5	—	Sept. 22, 1926
>300	—	92	100	180	Nov. 24, 1926
>250	—	95	65	—	Dec. 13, 1926
190	—	—	68	—	Dec. 31, 1926
100	—	86	55	66	Feb. 15, 1927

Table XLVIII. Milligrams of phosphate as P_2O_5 per cubic metre in the surface water at varying distances off shore (Atkins).

In Plymouth Sound	2 miles S. 21° W. (true) from W. Plymouth break-water	Off Eddy-stone Light-house	19 miles S. 21° W. (true) from Plymouth break-water	56½ miles S. 21° W. (true) from Plymouth break-water	Date
23	18	9	0	—	Sept. 13, 1923
12	17	15	22	—	Oct. 15, 1923
23	32	36	38	—	Jan. 2, 1924
35	30	35	32	31	Feb. 15, 1924
3	2	3	2½	—	June 17, 1924
3	2	0	2½	4*	July 9, 1924
13½	8	1½	1½	—	Aug. 7, 1924
28	28	15	12	—	Sept. 3, 1924
19	21	33	14	12	Nov. 12, 1924
35	35	33	32	—	Dec. 9, 1924

* At 5 metres.

The conditions leading to increased fertility in areas of moderate depth, as opposed to the areas of the ocean extending to a depth of several hundred metres or more, may be summarised as follows:

(1) The nitrate and phosphate re-formed at the bottom are carried into the upper layers in the vertical circulation of the water brought about by tidal streams and by convection currents due to cooling of the surface in winter.

(2) Such areas are liable to incursions of water, rich in phosphate and nitrate, rising from deep strata in the oceans.

(3) Regeneration of phosphate and nitrate takes place more quickly in the bottom water at moderate depths where it is not so cold as in the depths of the open ocean.

(4) Where the relatively shallow area is a coastal one, some nutrient salts are brought down in the rivers and streams.

(5) Where the area is a coastal one, strong off-shore winds will give rise to upwelling of deep water.

(6) The protozoa, bacteria and 'scavengers' living on dead organisms at the bottom and the purely bottom-living animals deriving nourishment from them are within the feeding range of many other animals in shallow areas, whether coastal or over an

isolated bank; whereas in deep ocean areas they are far distant from the fauna inhabiting moderate depths. This probably adds greatly to the fertility of the shallow Danish fiords whose bottoms are covered with weed and largely accounts for the great fertility of the shallow Egyptian lakes.

It is of speculative interest to consider factors affecting the relative fertility of areas of moderate depth such as 60 to 100 metres in the tropics and in temperate latitudes, neither being subject to incursions of deep water from the ocean. This is a condition hardly likely to be met with in nature, but such incursions would so modify the potential fertility that they need to be considered separately.

In both positions isothermal conditions from top to bottom tend to be set up by the action of tidal streams and wave motion, and such will be readily maintained in the tropics where the annual range of temperature at the surface is very small.

Table XLIX. Mean yearly range of surface temperature (Schott).

Equator	10° N.	20° N.	30° N.	40° N.	50° N.
2·3° C.	2·2° C.	3·6° C.	6·7° C.	10·2° C.	8·4° C.

The greater surface evaporation in the tropics will be an additional aid to vertical circulation. It follows that, owing to the small range in temperature, the water of such an area in the tropics tends to remain open to vertical circulation throughout the year, whereas in temperate latitudes it tends to be closed to vertical circulation during the summer months owing to layering.

In the tropics light penetrates to greater depths in sufficient quantity for the growth of phytoplankton, so the phosphate and nitrate regenerated at the bottom at the same depth in the tropics and in temperate latitudes respectively will have a lesser distance to rise in the tropic water before becoming available for plant growth. The total annual solar radiation is also greater in low than in high latitudes, although the total daily radiation during the long day at midsummer in high latitudes exceeds that at the equator (p. 135).

Probably the bacterial regeneration will be nearly twice as rapid in bottom water at 20° as at 10° C.

These factors all tend towards greater potential fertility of such an area in the tropics, but there is yet another factor of considerable magnitude which tends the other way. In the warmer waters all life proceeds with increased dissipation of energy owing to the greater rate of metabolism and this is not counterbalanced by the similarly increased rate of photosynthesis of the phytoplankton, because their growth is limited by the phosphate and nitrate supply. Taking the simple case of the loss and gain of combined carbon at two different temperatures, such as 10° and 20° C., by a mixed community of plants and animals, both will respire at a greater rate at the higher temperature and lose more carbon dioxide from the breakdown of their carbohydrates and fats. From laboratory experiments when the same animal is subjected to a more or less sudden change in temperature, the rate of loss of carbon dioxide is found to be almost doubled for a rise of 10° for such animals and plants as are naturally adapted for life within the experimental range of temperature. On the other side of the balance sheet, the gain in combined carbon owing to photosynthesis of carbon dioxide by the plants is also found by experiment to be about twice as rapid, when the plants have a sufficient supply of phosphate and nitrate. It may happen that these nutrient salts are re-formed from dead organisms on the bottom twice as rapidly at 20° as at 10° C., but whether they are able to reach the upper layers when photosynthesis proceeds in half the time taken at 10° is quite another matter. The case is still further complicated by the obvious criticism that it is not the same community which lives in tropical and temperate areas, and we do not actually know that an equal dry weight, exclusive of skeletons, of a tropic community will lose carbon dioxide in respiration at 20° twice as fast as the same dry weight of a temperate community at 10°.

FERTILITY AND LATITUDE IN DEEP OCEANS

The question next arises whether the average annual population is greater in the open oceans in higher latitudes where it is concentrated into the sunny months than in the tropical and subtropical zones with ample light throughout the year. The comparative estimations away from coastal banks have been made by plankton expeditions working in higher latitudes during the summer months only. Indeed to carry out the work throughout a North Atlantic winter with the necessarily delicate nets would be extremely difficult.

Considering Lohmann's comparative values given on p. 8, if the population in 40–45° N. remained in the same order of magnitude for the whole summer of six months as that found during the *Deutschland's* passage through these latitudes during May, the average annual population per litre from surface to 400 metres would be 3000 phytoplankton organisms. This value assumes that

Table L. Mean density of population of phytoplankton organisms per litre.

Latitude ...	50°–40° N.	40°–30° N.	30°–20° N.	20°–10° N.	10° N.	Equator
As found in ...	May 6000	June 2000	June 600	500	600	—
Surmised mean value for the whole year ...	3000	1000	600	500	600	—

phytoplankton is entirely absent during the remaining six months, whereas it is not absent, only much diminished. This value is greatly in excess of the tropical and subtropical populations. If a minimum annual population is calculated for 30–40° N. in the same, but in this case even more drastic manner, the value of 1000 organisms per litre is nearly double the population in the tropical and subtropical regions, which may presumably be taken as the mean annual population, since there is no sunless winter. The criticism might be raised that the population of 6000 per litre during May in 50–40° N. will not remain at this high value for so long as six months. If the average for the summer six months is only 2000 per litre, the average for the year is still twice the population found in the tropics.

A tentative explanation of the variation in fertility of the open ocean with latitude can be advanced. It has already been mentioned that an upwelling of water takes place in the equatorial region; this would account for a lesser amount of phytoplankton occurring between 10° and 20° N., as observed by Lohmann, and not between the equator and 10° N. On passing northward from the region 10–20° N. the annual production increases; at the same time the increase in density of the water with depth becomes less and less marked. Concurrently, as the severity of winter increases, convection currents penetrate deeper into the water below, rich in phosphate and nitrate, so that each spring not only does the phytoplankton have a greater store of those salts in the upper layers to draw upon, but the upper layers tend to remain in more open 'communication' with the rich depths for a longer period after

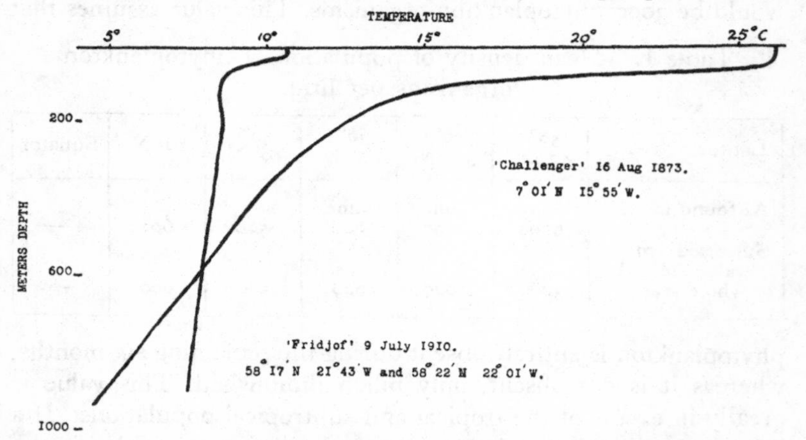

Fig. 61. Temperature gradient during the summer in equatorial and temperate latitudes of the North Atlantic.

the vernal onset of phytoplankton growth. Again it is apparent from Fig. 61 that both the depth of the epithalassa and the difference in density (which roughly follows the temperature) between epi- and hypothalassa is less in higher latitudes. Consequently the amount of nutrient salts brought annually into the upper layers by vertical mixing in the winter will be greater, since they have a lesser vertical distance to travel. Added to this is the effect of colder conditions upon the animals themselves, the more slowly they dissipate energy in the processes of metabolism, the greater is the number which can be supported upon a fixed ration of energy. That is to say the same annual production of phytoplankton will support a heavier animal population in cold than in tropic seas.

REGENERATION OF PHOSPHATES AND NITRATES

The regeneration of phosphates and nitrates from dead organisms is a process of interest, being a link in the chain of events which

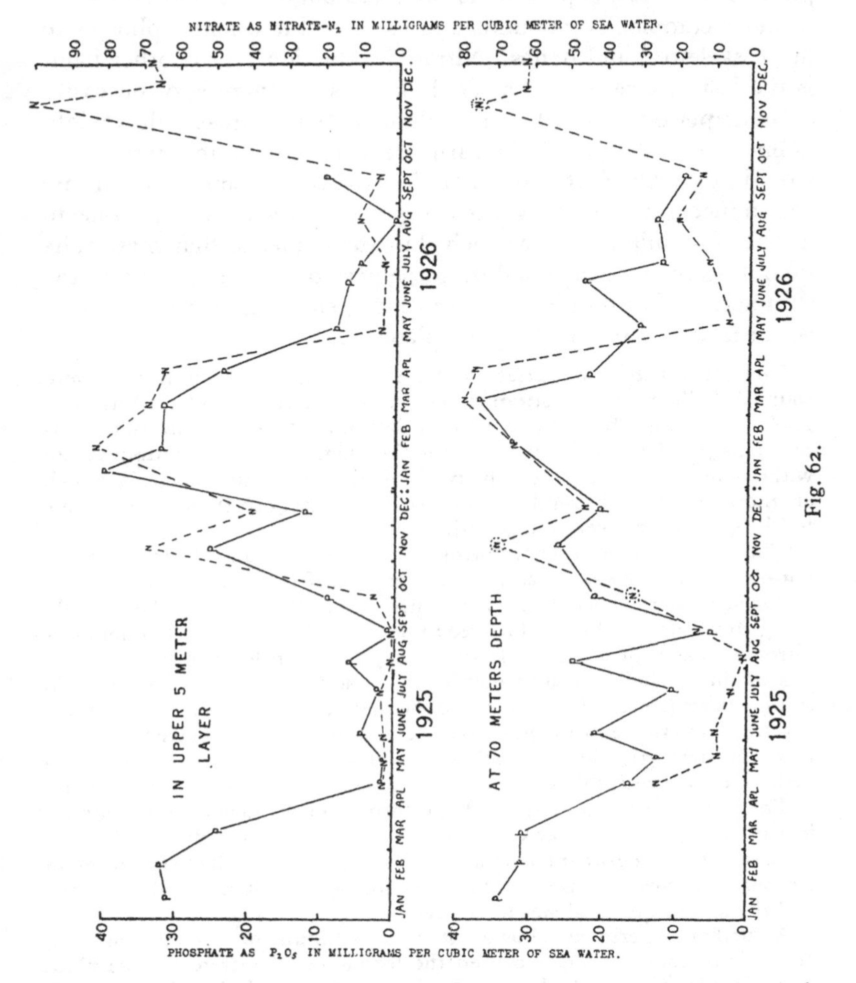

Fig. 62.

allows the continuation of marine life. Fig. 62 showing the annual variation of phosphate and nitrate in the English Channel indicates that during the summer months phosphate tends to be regenerated more rapidly than nitrate. As stated by Atkins ((4) 1926) the probable explanation is that the regeneration of phosphate is for the most part simply the splitting up of organic compounds in which the phosphate radicle is present as such, although a portion is present in more complex compounds and as insoluble calcium phosphate in exoskeletons and bones. Nitrate production on the other hand is the last of a series of bacterial changes, so there is of necessity a latent period while the protoplasm is being broken down into ammonium salts which in turn are converted into nitrates. It would appear at first sight that the complete change into nitrate was unnecessary, since various vegetative organisms are able to utilise ammonium salts as such, but the evidence that these salts do not decrease in the sea during the period of plant activity in the summer, as do phosphates and nitrates, suggests that ammonium is not utilised directly by phytoplankton.

Bacteria which form nitrate and nitrites from ammonium salts were found by Thomson in bottom deposits, derived from land washings, in Kiel Fiord, and by Issatchenko in bottom deposits from numerous positions (5). The latter remarks that the character of the bottom is not without influence upon the density of the nitrate-forming bacteria, which are more numerous in sand or bottoms rich in shellfish than in clay, while in black mud they were not found.

The writer found nitrate-forming bacteria in water from near the bottom in the English Channel, but not in surface water ((6) 1926).

Surface water collected 22 miles south-west of Plymouth on August 5th, 1925, to which had been added 200 mg. per cubic metre of ammonium-nitrogen, was kept in the dark for six days at room temperature. At the end of this period no colour developed on adding reduced strychnine in sulphuric acid. Water from a depth of 69 metres, collected on the same day and treated in the same manner, was kept in the dark together with a control to which mercuric chloride had been added. After six days this gave a definite colour with the reagent, while no colour was given by the control.

The depth of colour indicated the formation of nitrate to an amount in the order of 7 mg. per cubic metre. No nitrite was detectable. On repeating the experiment and keeping for eight days, likewise at room temperature, nitrate was again found in the water, but not in the control, and the amount formed was of the same order.

A further experiment showed that the addition of detritus collected about four miles off-shore caused the formation of nitrate to take place in surface water to which a trace of ammonium salt had been added, whereas without the detritus no nitrate was formed. The controls in which bacterial action was stopped by mercuric chloride showed no development of nitrate.

FLUCTUATIONS IN FERTILITY

Having dealt with the factors which give rise to a difference in density of population between relatively shallow and deep ocean areas, between relatively shallow areas in the tropics and in higher latitudes, and between deep ocean areas in different latitudes, it remains to discuss the more complicated question of the causes giving rise to fluctuations in the productivity of particular areas from year to year. The fluctuations in the density of *particular species*, especially of fish, have been a matter of active interest and the subject of much fishery research during the past twenty years. From this research several physical factors, such as temperature, currents or sunshine, have been demonstrated in certain instances to be the probable causes of fluctuations in the density of particular fish in certain areas; several such instances were described in the first chapter. Our present knowledge indicates some of the factors upon which fluctuations of the *total population* depend, and although the evidence is scanty and any conclusions drawn from it necessarily hypothetical, it is discussed here because fluctuations in the total population are probably linked up with the economically important fluctuations of particular fish. Any light upon the former may eventually be of assistance in the investigations, always complicated and often costly, which are being prosecuted by nearly all the larger maritime nations concerning fluctuations in the fisheries.

In temperate regions production of phytoplankton beyond the normal for a limited period during the summer months may not lead to increased production of animal life except on the bottom, since Lohmann's estimates in Kiel Bay indicate a surplus of phytoplankton throughout the summer; but an increased rain of vegetable matter falling on the bottom means increased food for the protozoa and bacteria which break up the vegetable protoplasm, and these in due course provide more food for the bottom fauna.

From a change in physical conditions bringing about an *extension* of the period in which there is an ample or more than ample supply of phytoplankton for maximum growth of the animals present, an increased annual production may be expected both of plants and of animals throughout the water strata and on the bottom.

In either of these events an increase takes place of both plants and animals. There are no observations as yet to confirm each of these two premises, nor is it likely that each separately is capable of confirmation, because the length of time during which a vigorous growth of phytoplankton takes place is different each year and the composition of the plankton is continually changing. Further, a quantitative estimation of the average annual population of plants and animals is wellnigh impossible over a sufficiently wide area to get representative values, and such an estimation would be necessary. All that can be stated from observation is the broad generalisation that in those areas where the phytoplankton is abundant animals are abundant, and *vice versa*.

However, if it is conceded that each of these two premises is correct, it follows that fluctuations in both phytoplankton and marine animals will be brought about in a particular area by:

(*a*) Fluctuations in the inflow of oceanic water from such depths that it is rich in nutrient salts.

Where the inflow occurs in the winter, giving rise to a larger stock of nutrient salts at the beginning of the vernal onset of vegetative growth, it should lead to a greater production of phytoplankton during one year than another. Where the inflow occurs during the summer months it will likewise lead to fluctuations. The greater surplus of vegetable matter over the needs of the animals in the water strata falls and nourishes more protozoan food for the bottom fauna; in due course this increased production will lead to increased regeneration at the bottom.

A drift of water takes place from the Atlantic into the Norwegian and North Sea through the Faeroe-Shetland Channel and is known to fluctuate in amount from year to year. This water is rich in nutrient salts ((7) 1926).

Table LI. In deep Channel between Faeroe and Shetland Isles, July 6, 1925.

Depth	Phosphate as P_2O_5, mg. per cubic metre	Nitrate as nitrate $-N_2$, mg. per cubic metre
10	30	67
60	53	160
300	—	160
500	43	—
800	43	—
900	—	160
1000	58	160

Atkins's estimations of phosphate lead to the conclusion that "from the incomplete data available for the region of the Faeroe-Shetland Channel it appears that the water there was considerably richer in phosphate during the summer of 1925 than it was during 1924" ((4) 1926). Hence fluctuations in the richness of the Atlantic water flooding into the Norwegian and North Sea may occur, in addition to the known fluctuations in its quantity.

The waters of the northern part of the North Sea are in general richer in phosphate than the waters of the English Channel, where the maximum occurring in the winter rarely exceeds 40 mg. P_2O_5 per cubic metre. There is some evidence to show that the nutrient salts in the upper layers of the deep water off Norway are replenished during the summer. At the two positions, distant only ten miles, shown in the following table, the observations indicate very little diminution in the phosphate in the upper layers between March and August, whereas at other positions such as off Plymouth, in the North Sea generally and off the Irish coast, a very marked diminution occurs during this interval.

Table LII.

	March 22, 1924 Position: lat. 57° 47′ N., long. 6° 19′ E.	August 30, 1924 Position: lat. 57° 51′ N., long. 6° 39′ E.	
Metres	Phosphate as P_2O_5, mg. per cubic metre	Phosphate as P_2O_5, mg. per cubic metre	° C.
0	22	20	16·14
10	—	—	16·04
20	28	16	9·89
50	31	28	6·26
75	—	27	6·32
100	34	27	6·12
150	37	39	5·97
200	37	30	5·45
250	37·5	28	5·38
280	39	—	—
300	—	42	5·19
300	—	42	5·19
350	—	26	5·10

From the distribution of temperature, the position is not one in which vertical mixing is sufficient to break down layering in the summer, so no material quantity of phosphate is likely to have been brought up by mixing from the water below, once layering had set in.

That fluctuations occur in the amount of nutrient salts brought into the North Sea area from year to year, obtains confirmation from the greater phosphate content of North Sea water in May 1925 compared with roughly the same season in 1924, as observed by Atkins ((4) 1926).

Further evidence of fluctuations due to inflow of oceanic water has been obtained in the mouth of the English Channel. In 1904 and 1921 very highly saline water (over 35·4 °/oo) entered and its passage westward could be clearly traced; in other years the distribution of salinity, the movement of bottom trailing drift bottles and the flow of water through Dover Straits have indicated that a current runs eastward past Ushant and eventually into the North Sea—it is not constant and regular, but rather in the nature of an intermittent current of variable velocity, averaging at times 1 to 1½ miles a day. At times it is probably non-existent; at times the movement into the North Sea is actually reversed by easterly winds in the Dover Straits neighbourhood.

Fig. 63 shows the course of this eastward drift and the position of three 'stations,' marked E_1, E_2 and E_3, where observations have been made at depths from top to bottom over a number of years.

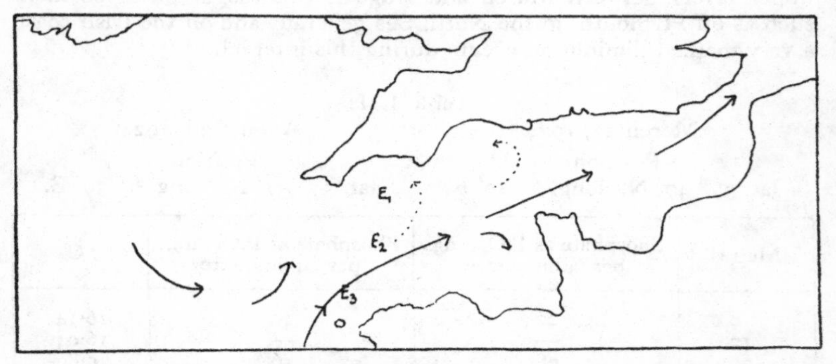

Fig. 63. Track of the intermittent drift up the English Channel.

Inflowing oceanic water has an apparent effect upon the phosphate content at the position marked E_3, which lies in the track of this intermittent current. During 1923 the water here was of low salinity and there was no inflow of oceanic water such as would be indicated by a sharp rise in salinity between consecutive observations which were made in April, May, July and November. At these times there was less than 20 mg. of phosphate reckoned as P_2O_5 per cubic metre.

During the summer of 1924 the movement of bottom trailing drift bottles in the Channel and the variable salinity found at E_3 indicated some movement of water up channel. In May and November the phosphate at E_3 was 20–25 mg., as P_2O_5, per cubic metre. During the middle of the summer, in July, it was only 14 mg. in round figures, having experienced a temporary fall probably due to consumption by phytoplankton.

During the summer of 1925 the phosphate remained at around 25 mg. P_2O_5 per cubic metre, the spring and summer consumption of phytoplankton being replaced.

In the early summer of 1926 an actual increase in phosphate took place, the water mass present in March having been replaced in May by a more saline water, richer in phosphate but not richer in nitrate. The increase took place during a period when consumption consistently exceeds regeneration in other areas, and was much greater than would be brought about by regeneration.

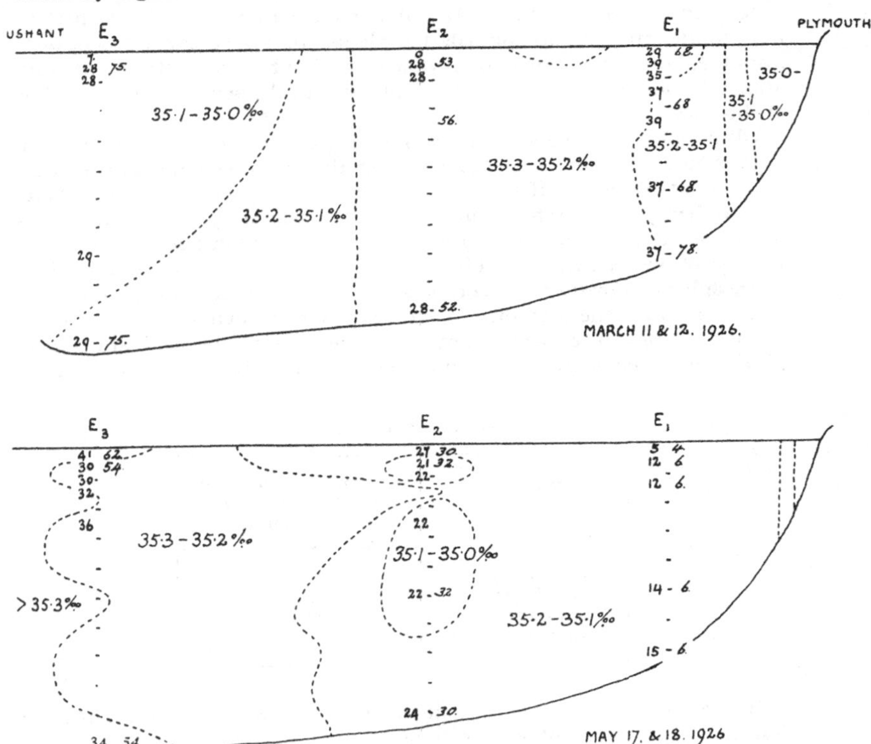

Fig. 64. Section across the English Channel between Plymouth and Ushant showing the distribution of salinity (°/oo) and of phosphate (upright figures show milligrams P_2O_5 per cubic metre) and of nitrate (slanting figures show milligrams nitrate-N_2 per cubic metre) as observed in March and May 1926.

No regular seasonal variation in phosphate takes place off Ushant, owing to the water mass being frequently replaced by others containing different amounts of this salt.

At Station E_1 which lies outside the track of the east-going current, as shown in Fig. 63, and is little affected by it during most years, a regular annual variation takes place in both the phosphate and the nitrate content.

During the winter when phytoplankton is sparse and its rate of consumption at a minimum, the phosphate and nitrate contents are at their greatest; during the spring outburst of diatoms the rate of consumption exceeds the rate of regeneration; through the summer the rates remain more or less level, a little more regeneration than consumption occurring occasionally and a little greater consumption than regeneration at other times. In the autumn and early winter regeneration from the corpses of the summer crop of both animals and plants greatly exceeds consumption, until finally, when the maximum values of phosphate and nitrate are again attained, the round of events is completed and the same state attained as at the start.

Observations for the years 1925 and 1926 are shown in Fig. 62, p. 179.

It is not suggested that the water in this E_1 area remains the same throughout the year. If it did the same minimum value for phosphate, and probably for nitrate, would occur each winter, provided there was no out of the ordinary winter growth of phytoplankton due to an unusual amount of winter sunshine. The variation in the winter maximum values of phosphate indicates that a certain small amount of nutrient salt normally enters or leaves the area during the year, but the extent to which these variations are due to water movement and the extent they are due to differences in consumption during December and January is not clear.

Table LIII.

Year (winter)	Winter maximum value of phosphate observed in mg. per cu. m.
1923/1924	37
1924/1925	32
1925/1926	40
1926/1927	—

Reference to Fig. 65 shows that the winter sunshine in 1924–1925 was less than during the other two winters.

Changes in salinity which occur at Station E_1 indicate some water movement having taken place between most monthly sets of observations, and it is concluded that the water masses which move in have themselves come from a similar area where they were subjected to similar conditions of plant growth. During exceptional years, such as 1903–1905 and 1921, this was certainly not the case, for then water of really high salinity, clearly of oceanic origin, entered the E_1 area.

(*b*) Physical conditions bringing about an extension of the period in which there is an ample or more than ample supply of phytoplankton for maximum growth of the animals present.

There is very definite evidence that the vernal onset of phytoplankton growth occurs earlier in some years than in others, and that it is well advanced when the daily sunshine exceeds about three hours. On inspection of Fig. 65, showing the daily sunshine in the area about Station E_1, it is clear that in 1926 and 1923 the spring sunshine occurred later in the year than in 1925 and 1924, the latter being the earliest of the four years

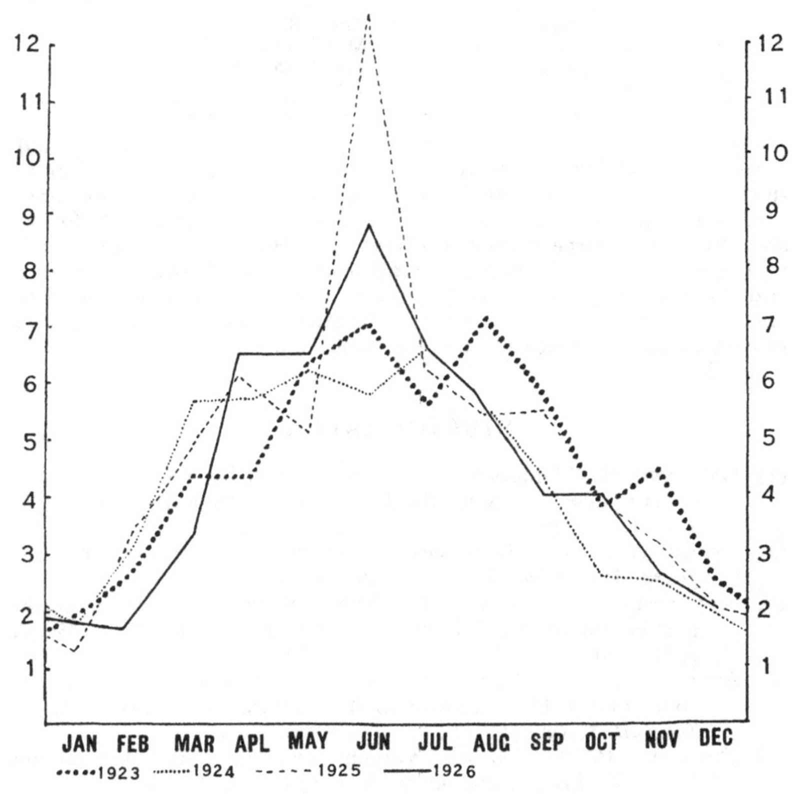

Fig. 65. Seasonal variation of the daily hours of sunshine 1923–1926. Average value of monthly means for Plymouth and Falmouth.

recorded. Directly in keeping with this are the dates at which the growth of phytoplankton had consumed phosphate in amount to reduce its concentration in the water from the winter maximum value to 20 mg. P_2O_5 per cubic metre (7).

It does not seem possible to gain any definite evidence concerning the end of the period of ample plant life from the nutrient salt content of the

Table LIV.

Year	Date at which phosphate in surface water at E_1 is reduced by plant growth to 20 mg. P_2O_5 per cu. m.
1923	April 28
1926	March 26
1925	March 20
1924	Feb. 28

water in the early winter. A very marked drop in both phosphate and nitrate occurred between November 11 and December 11, 1925, clearly shown in Fig. 62. This period was not in any way remarkable for more sunshine than normally occurs, but the autumn was remarkable for an abundance of a large diatom, *Rhizosolenia robusta*, in the area, to the probable agency of which a fall in dissolved silicate between October 1 and November 11 is attributed ((8) 1926). Whether this drop in phosphate and nitrate was due to an extension of the usual period of excess vegetative growth or to a temporary inflow of richer water in November is problematical.

BIBLIOGRAPHY

(1) LOHMANN, H. "Untersuchungen zur Feststellung des Vollständigen Gehalts des Meeres an Plankton." *Wiss. Meeresuntersuchungen*, **10**. Kiel, 1908.

(2) ATKINS, W. R. G. "Hydrogen-Ion Concentration in Sea Water." III. *Journ. Mar. Biol. Assoc.* **13**, 437–46. 1924.

(3) NATHANSOHN, A. "Ueber die Bedeutung vertikaler Wasserbewegungen für die Produktion des Planktons im Meere." *Abhand. Königl. sächs. Gesells. der Wissensch. Leipzig*, **39**, 3. 1906.

(4) ATKINS, W. R. G. "A Quantitative Consideration of some Factors concerned in Plant Growth in Water." Part II. *Journ. du Cons. Internat.* **1**, 197–226. 1926.

(5) THOMSON, P. "Ueber das Vorkommen von Nitrobakterien im Meere." *Wiss. Meeresuntersuchungen*, **11**. Kiel, 1910.

ISSATCHENKO, B. "Sur la Nitrification dans les Mers." *Compt. Rend. Acad. Sci. Paris*, **182**, 185–86. 1926.

(6) HARVEY, H. W. "Nitrate in the Sea." *Journ. Mar. Biol. Assoc.* **14**, 71–88. 1926.

(7) ATKINS, W. R. G. "The Phosphate Content of Sea Water." III. *Journ. Mar. Biol. Assoc.* **14**, 447–68. 1926.

(8) —— "Seasonal Changes in the Silica Content of the Water in Relation to the Phytoplankton." *Journ. Mar. Biol. Assoc.* **14**, 89–100. 1926.

INDEX

Absorption coefficient of light, 155, 160–162
  of carbon dioxide, 65
Acclimatisation to temperature, 24
Addition of alkali to sea water, 75
  of acid to sea water, 76
Adeney, W. E., 62, 82
Adiabatic heating, 108, 151
Adjoining seas, 149
Aeration, effect of, on albuminoid ammonia, 49
  of sea water, 61
  rate of, 61
Aitken, J., 116, 131
Albuminoid ammonia, 54
Algae, blue-green, 8
  fixed, 5, 6; adaptation to low temperature, 17; effect of *p*H on, 73, 80
  planktonic, *see* Phytoplankton
Alkali reserve, 64
Allen, E. J., 18, 26, 33, 34
Aluminium in sea water, 51
Ammonium salts, 14, 48
Ångström, A., 135, 154
Animal plankton, 19
Animals, marine, 19–32
  effect of currents on, 20; of light on, 25–27; of salinity on, 28; of temperature on, 21–24
Anticyclonic wind, pool-making by, 109
Aquaria, plunger jar, 57
  storage of water for, 49
  survival of organisms in, 31, 32
Archimedean forces in the sea, 99, 110
Arsenates, replacing phosphates as nutrient salts, 51
Arsenic in sea water, shellfish, etc., 51
Atkins, W. R. G., 10, 32, 40, 41, 42, 51, 73, 74, 81, 82, 83, 140, 154, 163, 180, 183, 188
Atlantic, North, 4, 169
  currents in, 114
  densely populated areas in, 170
  salinity distribution in, 117
  temperature distribution in, 145–148
Atlantic Stream, 116, 149
Atmosphere, exchange of carbon dioxide with, 63, 64
  humidity of, and evaporation, 138
Azotobacter, 14

Bacteria, 31, 167
  and nature of bottom, 180
  and sunshine, 31
  denitrifying, 11, 12, 172
  in lakes, 13
  nitrate-forming, 14, 180
  nitrogen-fixing, 13, 14
  pathogenic, 31
Baltic, 56, 149
Barcroft, J., 82
Barents Sea, 117
Barium in sea water, 51
Barometric pressure and height of tide, 106
Bases in sea water, 63; *see also* Excess base
Bauer, E., 11, 32
Bertel, R., 31, 35
Bicarbonates, 63, 67
Bidder, G. P., 97
Birge, E. A., 154
Bjerknes' Circulation Theory, 122–130
Black Sea, 56, 151
Blackman, F. F., 33
Blevgard, H., 31, 34
Bohr, invasion and evasion coefficients, 61
Boron, in sea water, 51
Bottazzi, F., 34
Bottom, effect of nature of, on bacterial fauna, 180
Bottom fauna, 31, 174
Bowman, A., 119, 132
Boysen Jensen, 34
Brandt, K., 11, 12, 33, 43, 44, 48, 49, 50, 81, 83
Brennecke, W., 55, 82, 116
Briggs, G. E., 33
Bristol, M. and Page, H., 13, 33
Bromides, 36
Browne, E. T., 57, 82
Bruce, J. R., 69, 82
Buffer solutions, 77

Cabelling, 113
Caesium in sea water, 51
Calcium in sea water, 36
  precipitation of, 75
Carbon dioxide, absorption coefficient, 65
  in sea water, 63–76; *Krogh's* determinations, 63; relation to excess base and *p*H, 68; total, measurement of, 64
  tension in sea water, 64; measurement of, 64; relation to *p*H, 70
Carbonates in sea water, 36, 63, 67
Carbonic acid, undissociated, in sea water, 65
Carruthers, J. N., 98, 130, 131
Challenger, H.M.S., 1
Chlorides in sea water, 36
Clarke, F. W., 36, 81
Coastal areas, fertility of, 173
Cobalt in marine organisms, 29, 51
Coefficient, absorption of CO_2, 65; of light, 155
  invasion and evasion of gases, 61
  temperature, 24
  virtual, of thermal conductivity, 134; of viscosity, 102
'Cold wall,' 113
Colour of the sea, 155–163
Compressibility of sea water, 108
Copper in sea water, 51
Cresswell, M., 131
Current meters, 98
Currents, 19, 84, 94–132
  anticyclonic, 109
  ascending, 104–109
  carrying plankton, 19, 20, 119
  cyclonic, 111
  direction of wind-produced, 101, 103
  effect of horizontal, on fauna, 20, 21, 119
  effect of rotation of earth on, 100
  fluctuations in, 2, 118, 152, 182–186
  in English Channel, 184
  in entrance to Baltic, 150
  in North Atlantic, 114
  in Straits of Gibraltar, 112
  produced by Archimedean forces, 110–114; by ice melting, 112; by wind, 101–109
Cycle of life in the sea, 167

Dakin, W. J., 34
Datum level, 89
Dead reckoning, 97
Denigès, G., 40, 44
Density of population, cycle of changes regulating, 167
  effect of horizontal and vertical currents on, 19–21; of river water on, 173; of temperature and rate of metabolism on, 176
  fluctuations in, 25–27, 181–188
  in various latitudes, 8, 176
Density of sea water, change of, with temperature, 141
  in situ, 39
  measurement of, 38
  relation to salinity, 39
Detritus as food of bottom fauna, 30, 174
Diatoms, 7–10, 18
  chromatophores of, and latitude, 14
  distribution with depth, 9
Dinoflagellates, 7
Discontinuity layer, 107, 140
Diurnal variation, in temperature, 145
  in dissolved oxygen, 54
Dogger Bank, 20
Domogalla, 13, 32, 54
Doodson, A. T., 131
Drew, C. H., 32, 44
Drift bottles, 97, 103
Duclaux and Jeantet, 163

Ekman, V., 39, 81, 101, 108, 131
Epithalassa, 140
Evaporation, 136–139
Excess base, 64
  and total CO_2, 68
  measurement of, 66

Feitel, R., 11, 32
Fertility, 164
  and latitude, 176
Filtration of sea water, 54
Fisheries, fluctuations in, 2, 5, 20, 150
  value of, 5
Flatfish, fluctuations in numbers, 20
'Flood and Ebb,' 95
Fluctuations, in currents, 2, 20, 118, 152, 182–186
  in productivity, 181
  in sea temperature, 2, 118, 138, 152
Fluorine in sea water, 51
Food chain in the sea, 167

Ford, E., 20, 33
Fox, C. J. J., 59, 82, 83
'Fram,' drift of, 103

Gaarder, T., 35, 73, 82
Gail, 80, 163
Galapagos Islands, 105
Garstang, W., 31, 131
Giral, J., 37
Gold in sea water, 51
Gran, H., 11, 12, 15, 29, 32, 35, 172
Gran, H. and Ruud, B., 35, 53, 83
Grand Banks, 113
Grein, K., 161
Gulf Stream, 104
  fluctuations in, 118
  off Grand Banks, 113

Haas, A. R., 83
Haemocyanin, 51, 58
Haemoglobin, 58
Hardy, A. C., 119, 132
Harris' theory of tides, 93
Harvey, H. W., 35, 44, 81, 82, 118, 132, 138, 154, 180
Hastings and Sendray, 65, 82
Heat, absorption of, 134
  conduction of, in the sea, 134
  gain and loss by the sea, 133–138
  lost by evaporation, 138
  specific, of sea water, 113
Helland-Hansen, B., 82, 116, 122, 131, 132, 152, 154
Herdman, G., 32
Hesselburg and Sverdrup, 124, 131, 132
Hjort, J., 3, 26
Hogben, L., 22, 82
Hunt, O. D., 31, 34
Hydrogen ion concentration and light, 72
  and *p*H, 69
  biological effects of, 79
  determination of, 76
  distribution in the oceans, 72–74
  effect of temperature on, 71
  relation to total CO_2 and excess base, 68; to CO_2 tension and excess base, 70
  showing excess respiration over consumption of CO_2, 72, 73
Hydrographical tables, *Knudsen's*, 37, 39
Hypothalassa, 140

Ice, currents produced by melting of, 113
Ice Patrol, International Service, 4, 39, 83, 129
Indicators, effect of proteins on, 77; of salts on, 77; of temperature on, 79
  range of common, 79
Interferometer, 39
Iodine in sea water and algae, 51
Ionic product, K_W, 69
Iron in sea water, 51
Irvine, L., 76, 82, 83
Isosteres, 124
Issatchenko, B., 180, 188

Jacobsen, J., 54, 82
Jeffreys, H., 131
Johansen, A. G., 22, 34, 150, 154

Kanitz, A., 33
Keding, M., 14, 33
Keutner, J., 14, 33
Klugh, A. B., 16, 34, 162, 163
Kneip, H., 16, 33
Knudsen, M., 36, 37, 81, 162
Krikensky, J., 34
Krogh, A., 22, 34, 61, 63, 82
Krummel, O., 2, 105, 131

Labrador Current, 114
Lakes, amino acids in, 13
  organic matter in, 13
  temperature of, 134
Lead in marine organisms, 52
Legendre, R., 18, 54, 79, 80, 83
Light, absorption in sea water, 155–163
  and mackerel, 27
  and phytoplankton, 15, 187
  effect of intensity on photosynthesis, 15, 16; of quality on photosynthesis, 16
  quality at different depths, 157
  ultra-violet, 31, 155, 160
Lithium in sea water, 51
Lohmann, H., 8, 32, 165, 177, 181, 188
Lütgens, R., 154

Macallum, A. B., 28, 34
Mackerel, and sunshine, 27
  and temperature of the Baltic, 150
Magnesium in sea water, 36
Manganese in sea water and algae, 52

Marmer, H. A., 130
Marshall, S. M. and Orr, A. P., 42, 50, 54, 73, 83
Matthaei, G., 17, 33
Matthews, D., 40, 81, 131
Mayer, A. G., 24, 34
McClendon, J. F., 43, 68, 70, 76, 77, 82, 83
McEwen, G. F., 138, 154
Mean sea level, 92, 118
Mediterranean, bacteria in, 31
  currents, 111
  dissolved organic matter, 49
  temperature of, 149
Memel Deep, 105
Merz, A., 116, 132
Metabolism, effect of temperature on, 17, 21–24, 176
  and density of population, 21
Microplankton, 7, 8
Moore, B., 10, 13, 32, 33
Movement along surfaces of equal density, 107

Nansen, F., 116, 122, 131, 132, 152, 154
Nathansohn, A., 12, 13, 168, 188
Nelson, E. W., 130
Nickel in molluscs, 29, 52
Nitrates, 11–14, 42–48, 168, 170, 179
  in English Channel, 45, 173, 179, 185
  in North Atlantic, 46, 47, 182
  in Plymouth Sound, 46, 173
  regeneration of, 13, 179
Nitrate-forming bacteria, 14, 180
Nitrites in sea water, 49
Nitrite-forming bacteria, 180
Nitrogen, dissolved, 55, 61
Nitzschia, phosphate used during multiplication of, 10
  culture of, 19
'Normal Water,' 37
North Atlantic, *see* Atlantic
North Sea, fluctuations in productivity of, 182
Norwegian Sea, 151, 182
Nutrient salts, and plant life, 11–13, 40–48, 81, 95, 168
  in river water, 173
  store in deep ocean water, 41, 45, 170

Off-shore and on-shore winds, effect of, 105–109, 129
Oil film and rate of solution of oxygen, 63
Organic matter dissolved in sea water, 52
  in lake water, 53
Orr, A. P., 50, 82; *see also Marshall and Orr*
Orton, J. H., 23, 34
Osmotic pressure, 28, 80
Osterhout and Haas, 33
Oxner and Knudsen, 81
Oxygen, and animal life, 21, 56–59
  dissolved, 54–63
  distribution in the oceans, 55, 150, 151
  estimation of, 59
  rate of solution of, 61–63
  solubility of, 60

Page, H., 33
Pantin, C. F. A., 25, 34, 82
Pape, C., 39
Peach and Drummond, 19
Pentosans, 31
Peridinians, 7
Periodicity, 152
Peters, R. H., 30, 34
Pettersson, O., 116, 122, 131, 132, 152
*p*H, definition of, 69
Phosphates, estimation of, 42
  in English Channel, 40, 174, 179, 185
  in North Atlantic, 41
  in North Sea, 182, 183
  in sea water, 11, 12, 40, 168, 171
  regeneration of, 178
  required by phytoplankton, 11
Photosynthesis, 14, 15
  effect of intensity of light, 15, 16; of lack of nutrient salts on, 14; of quality of light on, 16; of salinity on, 18; of temperature on, 17
Phytoplankton, 5, 7
  annual production, 10, 11, 164–188; and latitude, 176
  artificial culture, 18
  distribution, 7–9
  neritic and oceanic, 7
Pietenpol, W., 158, 163
Plaetzer, H., 17, 33
Plaice, eggs, rate of development of, 22
  growth on Dogger Bank, 20
Plunger jar, 57
Poole, H. H., 163

Potassium hydride cell, 159
Potassium salts in sea water, 36
Pouget and Chouchak, 40
Prideaux, E., 32
Pseudomonas calcis, 44
Pure water, absorption of light by, 156
  *p*H of, 69
Pütter's Theory, 29, 34

Raben, E., 40, 43, 49, 53
Race, tidal, 94
Radiation, outward from the sea, 136
  solar, 133–135
Radioactivity of sea water, 52
Range of temperature in the sea, 23, 144, 175
  of tide, 86, 89, 91, 92
Refractive index of sea water, 39
Regeneration of phosphates and nitrates, 179
Reinke, E., 13, 33
Respiration, animal, 21, 58
  plant, 15, 17
Respiratory pigments, 58
Roach, B., 30, 34
Roberts, C. H., 63, 83
Rotation of the earth, deflective force due to, 100
Rubidium in sea water, 52
Russell, F. S., 27, 34

Salinity, 36
  distribution in North Atlantic, 117
  effect on distribution of fauna, 28
  estimation of, 37, 38, 39
  relation to chlorine content, 36; to density, 39; to temperature, 28, 39
Salt error, 77, 79
Salts, absorbed by marine organisms, 29
  in body fluids of marine animals, 28
  in sea water, 36 et seq.
  see also Salinity
Sandström, J. W., 107, 121, 126, 131, 132
Sargasso Sea, colour of, 157
  temperature distribution, 110
Saunders, J. T., 71, 79, 83
Schmidt, J., 58, 82, 132
Schmidt, W., 154
Schott, G., 105, 131, 154
Schumacher, A., 39
Sea water, addition of acid to, 76; of alkali to, 75
Sea water as a medium for plant growth and animal life, 80
Secchi's disc, 156
Seiches, 122
Sensitivity to light, relative, 159
Shelford and Gail, 163
Silicon in sea water, 50
  estimation of, 50
Silver in sea water, 52
Smith, E. H., 132
Sodium salts in sea water, 36
Solar radiation, absorption in the sea, 134, 155
  relation to daily sunshine, 135
  variation with latitude, 135
  wave length of, 133
  see also under Light
Solubility, of calcium carbonate in sea water, 75
  of carbon dioxide in water, 65
  of oxygen in sea water, 62
Specific heat of sea water, 113
Specific volume, 124
Stationary waves, 94, 121
Stedman and Stedman, 82
Storage of sea water in dark, changes during, 49
Stowell, F. P., 49
Strontium forming skeleton of radiolarian, 29, 52
Submarine waves, 120
Sulphates in sea water, 36
Sulphuretted hydrogen, 151
Sund, O., 132
Sunshine, and dissolved phosphate, 171
  and growth of phytoplankton, 15
  and mackerel, 26
  and *p*H, 72
  and solar radiation, 135
  in English Channel, 187
Sverdrup, H., 131

Temperature, effect on breeding season, 23; on distribution of the fauna, 24; on metabolism, 21–25, 176, 178; on photosynthesis, 17; on rate of development of plaice eggs, 22
Temperature coefficient, 25
Temperature of air, Copenhagen, 150
  Norwegian coast, 152
Temperature of sea, distribution with depth, 139–144, 150, 151, 178; on approaching land, 142

Temperature of sea, diurnal variation, 145
 fluctuations from year to year, 2, 118, 138, 152
 influence of currents on, 118, 119, 145–148, 152; of evaporation on, 138
 of Arctic, 151
 of Baltic, 150
 of Black Sea, 151
 of Mediterranean, 149
 of North Atlantic, 146, 147, 148, 178
 of Norwegian Sea, 151
 of surface water, 146, 175
 range of, 23, 144; and latitude, 175
 redistribution of, in the sea, 134
Thalassiosira, culture of, 19
Theil and Strohecker, 65, 82
Thermocline, *see* Discontinuity layer
Thomson, P., 180, 188
Thoulet, J., 163
Tidal constants, 91
Tidal differences, 90
Tidal streams, atlases showing flood and ebb of, 96
 diurnal and semi-diurnal, 95
 effect of, on inshore temperature, 142
 rotary, 95
 velocity of, 94
Tidal zone, fertility of, 30
 flora of, 6
Tides, 84–94
 age of, 88, 92
 diurnal and semi-diurnal, 88
 effect of wind on, 106; of barometric pressure on, 106
 priming and lagging, 91
Tides, range of, 86, 89, 91, 92
 spring and neap, 87
 to find height at times between high and low water, 91, 93
Tornöe, H., 83
Transparency of sea water, 156

Ultra-violet light, absorption of, by sea water, 155, 160, 161
 effect on marine bacteria, 31

Vanadium in blood of marine animals, 29, 52
Vernon, H. M., 21, 49, 54, 82
Vertical mixing, 168

Walker and Cormack, 65, 82
Wann, 13
Warburg, E., 66, 82
Warburg, H. D., 86, 130
Wattenberg, H., 67, 171
Wegemann, G., 145, 154
Werenskiold, W., 96, 130
Witting, R., 103, 106, 131
Wind, currents produced by, 101–109; by change in direction with depth, 101
 effect on height of tide, 106
 force exerted by, 101
 off-shore and on-shore effect of, 105, 107, 108
Winkler's method of oxygen determination, 59
Wurmser, R., 16, 33
Wyville Thomson Ridge, 47, 48, 56

Zinc in sea water, 52
Zooplankton, 19

Printed in the United States
By Bookmasters